S0-AYT-510

BRITANNICA
**Mathematics
in
Context**

Looking at
an Angle

⊕Britannica

ENCYCLOPÆDIA BRITANNICA EDUCATIONAL CORPORATION

MAR 2 3 1999

CURRICULUM LIBRARY
GOSHEN COLLEGE LIBRARY
GOSHEN, INDIANA

Mathematics in Context is a comprehensive curriculum for the middle grades. It was developed in collaboration with the Wisconsin Center for Education Research, School of Education, University of Wisconsin–Madison and the Freudenthal Institute at the University of Utrecht, The Netherlands, with the support of National Science Foundation Grant No. 9054928.

National Science Foundation

Opinions expressed are those of the authors
and not necessarily those of the Foundation

Copyright © 1998
Encyclopædia Britannica Educational Corporation

All rights reserved.
Printed in the United States of America.

This work is protected under current U.S. copyright laws, and the performance, display, and other applicable uses of it are governed by those laws. Any uses not in conformity with the U.S. copyright statute are prohibited without our express written permission, including but not limited to duplication, adaptation, and transmission by television or other devices or processes. For more information regarding a license, write Encyclopædia Britannica Educational Corporation, 310 South Michigan Avenue, Chicago, Illinois 60604.

ISBN 0-7826-1534-1
1 2 3 4 5 WK 02 01 00 99 98

The *Mathematics in Context* Development Team

Mathematics in Context is a comprehensive curriculum for the middle grades. The National Science Foundation funded the National Center for Research in Mathematical Sciences Education at the University of Wisconsin–Madison to develop and field-test the materials from 1991 through 1996. The Freudenthal Institute at the University of Utrecht in The Netherlands, as a subcontractor, collaborated with the University of Wisconsin–Madison on the development of the curriculum.

The initial version of *Looking at an Angle* was developed by Els Feijs, Jan de Lange, and Martin van Reeuwijk. It was adapted for use in American schools by Mary S. Spence and Jonathan Brendefur.

National Center for Research in Mathematical Sciences Education Staff

Thomas A. Romberg
Director

Joan Daniels Pedro
Assistant to the Director

Gail Burrill
Coordinator
Field Test Materials

Margaret R. Meyer
Coordinator
Pilot Test Materials

Mary Ann Fix
Editorial Coordinator

Sherian Foster
Editorial Coordinator

James A. Middleton
Pilot Test Coordinator

Margaret A. Pligge
First Edition Coordinator

Project Staff

Jonathan Brendefur
Laura J. Brinker
James Browne
Jack Burrill
Rose Byrd
Peter Christiansen
Barbara Clarke
Doug Clarke
Beth R. Cole

Fae Dremock
Jasmina Milinkovic
Kay Schultz
Mary C. Shafer
Julia A. Shew
Aaron N. Simon
Marvin Smith
Stephanie Z. Smith
Mary S. Spence
Kathleen A. Steele

Freudenthal Institute Staff

Jan de Lange
Director

Els Feijs
Coordinator

Martin van Reeuwijk
Coordinator

Project Staff

Mieke Abels
Nina Boswinkel
Frans van Galen
Koeno Gravemeijer
Marja van den Heuvel-Panhuizen
Jan Auke de Jong
Vincent Jonker
Ronald Keijzer

Martin Kindt
Jansie Niehaus
Nanda Querelle
Anton Roodhardt
Leen Streefland
Adri Treffers
Monica Wijers
Astrid de Wild

Acknowledgments

Several school districts used and evaluated one or more versions of the materials: Ames Community School District, Ames, Iowa; Parkway School District, Chesterfield, Missouri; Stoughton Area School District, Stoughton, Wisconsin; Madison Metropolitan School District, Madison, Wisconsin; Milwaukee Public Schools, Milwaukee, Wisconsin; and Dodgeville School District, Dodgeville, Wisconsin. Two sites were involved in staff developments as well as formative evaluation of materials: Culver City, California, and Memphis, Tennessee. Two sites were developed through partnership with Encyclopædia Britannica Educational Corporation: Miami, Florida, and Puerto Rico. University Partnerships were developed with mathematics educators who worked with preservice teachers to familiarize them with the curriculum and to obtain their advice on the curriculum materials. The materials were also used at several other schools throughout the United States.

We at Encyclopædia Britannica Educational Corporation extend our thanks to all who had a part in making this program a success. Some of the participants instrumental in the program's development are as follows:

Allapattah Middle School
Miami, Florida
Nemtalla (Nikolai) Barakat

Ames Middle School
Ames, Iowa
Kathleen Coe
Judd Freeman
Gary W. Schnieder
Ronald H. Stromen
Lyn Terrill

Bellerive Elementary
Creve Coeur, Missouri
Judy Hetterscheidt
Donna Lohman
Gary Alan Nunn
Jakke Tchang

Brookline Public Schools
Brookline, Massachusetts
Rhonda K. Weinstein
Deborah Winkler

Cass Middle School
Milwaukee, Wisconsin
Tami Molenda
Kyle F. Witty

Central Middle School
Waukesha, Wisconsin
Nancy Reese

Craigmont Middle School
Memphis, Tennessee
Sharon G. Ritz
Mardest K. VanHooks

Crestwood Elementary
Madison, Wisconsin
Diane Hein
John Kalson

Culver City Middle School
Culver City, California
Marilyn Culbertson
Joel Evans
Joy Ellen Kitzmiller
Patricia R. O'Connor
Myrna Ann Perks, Ph.D.
David H. Sanchez
John Tobias
Kelley Wilcox

Cutler Ridge Middle School
Miami, Florida
Lorraine A. Valladares

Dodgeville Middle School
Dodgeville, Wisconsin
Jacqueline A. Kamps
Carol Wolf

Edwards Elementary
Ames, Iowa
Diana Schmidt

Fox Prairie Elementary
Stoughton, Wisconsin
Tony Hjelle

Grahamwood Elementary
Memphis, Tennessee
M. Lynn McGoff
Alberta Sullivan

Henry M. Flagler Elementary
Miami, Florida
Frances R. Harmon

Horning Middle School
Waukesha, Wisconsin
Connie J. Marose
Thomas F. Clark

Huegel Elementary
Madison, Wisconsin
Nancy Brill
Teri Hedges
Carol Murphy

Hutchison Middle School
Memphis, Tennessee
Maria M. Burke
Vicki Fisher
Nancy D. Robinson

Idlewild Elementary
Memphis, Tennessee
Linda Eller

Jefferson Elementary
Santa Ana, California
Lydia Romero-Cruz

Jefferson Middle School
Madison, Wisconsin
Jane A. Beebe
Catherine Buege
Linda Grimmer
John Grueneberg
Nancy Howard
Annette Porter
Stephen H. Sprague
Dan Takkunen
Michael J. Vena

Jesus Sanabria Cruz School
Yabucoa, Puerto Rico
Andreíta Santiago Serrano

John Muir Elementary School
Madison, Wisconsin
Julie D'Onofrio
Jane M. Allen-Jauch
Kent Wells

Kegonsa Elementary
Stoughton, Wisconsin
Mary Buchholz
Louisa Havlik
Joan Olsen
Dominic Weisse

Linwood Howe Elementary
Culver City, California
Sandra Checel
Ellen Thireos

Mitchell Elementary
Ames, Iowa
Henry Gray
Matt Ludwig

New School of Northern Virginia
Fairfax, Virginia
Denise Jones

Northwood Elementary
Ames, Iowa
Eleanor M. Thomas

Orchard Ridge Elementary
Madison, Wisconsin
Mary Paquette
Carrie Valentine

Parkway West Middle School
Chesterfield, Missouri
Elissa Aiken
Ann Brenner
Gail R. Smith

Ridgeway Elementary
Ridgeway, Wisconsin
Lois Powell
Florence M. Wasley

Roosevelt Elementary
Ames, Iowa
Linda A. Carver

Roosevelt Middle
Milwaukee, Wisconsin
Sandra Simmons

Ross Elementary
Creve Coeur, Missouri
Annette Isselhard
Sheldon B. Korklan
Victoria Linn
Kathy Stamer

St. Joseph's School
Dodgeville, Wisconsin
Rita Van Dyck
Sharon Wimer

St. Maarten Academy
St. Peters, St. Maarten, NA
Shareed Hussain

Sarah Scott Middle School
Milwaukee, Wisconsin
Kevin Haddon

Sawyer Elementary
Ames, Iowa
Karen Bush Hoiberg

Sennett Middle School
Madison, Wisconsin
Brenda Abitz
Lois Bell
Shawn M. Jacobs

Sholes Middle School
Milwaukee, Wisconsin
Chris Gardner
Ken Haddon

Stephens Elementary
Madison, Wisconsin
Katherine Hogan
Shirley M. Steinbach
Kathleen H. Vegter

Stoughton Middle School
Stoughton, Wisconsin
Sally Bertelson
Polly Goepfert
Jacqueline M. Harris
Penny Vodak

Toki Middle School
Madison, Wisconsin
Gail J. Anderson
Vicky Grice
Mary M. Ihlenfeldt
Steve Jernegan
Jim Leidel
Theresa Loehr
Maryann Stephenson
Barbara Takkunen
Carol Welsch

Trowbridge Elementary
Milwaukee, Wisconsin
Jacqueline A. Nowak

W. R. Thomas Middle School
Miami, Florida
Michael Paloger

Wooddale Elementary Middle School
Memphis, Tennessee
Velma Quinn Hodges
Jacqueline Marie Hunt

Yahara Elementary
Stoughton, Wisconsin
Mary Bennett
Kevin Wright

Site Coordinators

Mary L. Delagardelle—Ames Community Schools, Ames, Iowa

Dr. Hector Hirigoyen—Miami, Florida

Audrey Jackson—Parkway School District, Chesterfield, Missouri

Jorge M. López—Puerto Rico

Susan Militello—Memphis, Tennessee

Carol Pudlin—Culver City, California

Reviewers and Consultants

Michael N. Bleicher
Professor of Mathematics
University of Wisconsin–Madison
Madison, WI

Diane J. Briars
Mathematics Specialist
Pittsburgh Public Schools
Pittsburgh, PA

Donald Chambers
Director of Dissemination
University of Wisconsin–Madison
Madison, WI

Don W. Collins
Assistant Professor of Mathematics Education
Western Kentucky University
Bowling Green, KY

Joan Elder
Mathematics Consultant
Los Angeles Unified School District
Los Angeles, CA

Elizabeth Fennema
Professor of Curriculum and Instruction
University of Wisconsin–Madison
Madison, WI

Nancy N. Gates
University of Memphis
Memphis, TN

Jane Donnelly Gawronski
Superintendent
Escondido Union High School
Escondido, CA

M. Elizabeth Graue
Assistant Professor of Curriculum and Instruction
University of Wisconsin–Madison
Madison, WI

Jodean E. Grunow
Consultant
Wisconsin Department of Public Instruction
Madison, WI

John G. Harvey
Professor of Mathematics and Curriculum & Instruction
University of Wisconsin–Madison
Madison, WI

Simon Hellerstein
Professor of Mathematics
University of Wisconsin–Madison
Madison, WI

Elaine J. Hutchinson
Senior Lecturer
University of Wisconsin–Stevens Point
Stevens Point, WI

Richard A. Johnson
Professor of Statistics
University of Wisconsin–Madison
Madison, WI

James J. Kaput
Professor of Mathematics
University of Massachusetts–Dartmouth
Dartmouth, MA

Richard Lehrer
Professor of Educational Psychology
University of Wisconsin–Madison
Madison, WI

Richard Lesh
Professor of Mathematics
University of Massachusetts–Dartmouth
Dartmouth, MA

Mary M. Lindquist
Callaway Professor of Mathematics Education
Columbus College
Columbus, GA

Baudilio (Bob) Mora
Coordinator of Mathematics & Instructional Technology
Carrollton-Farmers Branch Independent School District
Carrollton, TX

Paul Trafton
Professor of Mathematics
University of Northern Iowa
Cedar Falls, IA

Norman L. Webb
Research Scientist
University of Wisconsin–Madison
Madison, WI

Paul H. Williams
Professor of Plant Pathology
University of Wisconsin–Madison
Madison, WI

Linda Dager Wilson
Assistant Professor
University of Delaware
Newark, DE

Robert L. Wilson
Professor of Mathematics
University of Wisconsin–Madison
Madison, WI

TABLE OF CONTENTS

BRITANNICA
Mathematics in Context

Dear Teacher,

Welcome! *Mathematics in Context* is designed to reflect the National Council of Teachers of Mathematics Standards for School Mathematics and to ground mathematical content in a variety of real-world contexts. Rather than relying on you to explain and demonstrate generalized definitions, rules, or algorithms, students investigate questions directly related to a particular context and construct mathematical understanding and meaning from that context.

The curriculum encompasses 10 units per grade level. *Looking at an Angle* is designed to be the third in the geometry strand for grade 7/8, but the unit also lends itself to independent use—to introduce students to calculating the steepness, or the ratio of height to distance, of a right triangle, and to finding the relationship between the angles and the lengths of the sides of a right triangle.

In addition to the Teacher Guide and Student Books, *Mathematics in Context* offers the following components that will inform and support your teaching:

- *Teacher Resource and Implementation Guide,* which provides an overview of the complete system, including program implementation, philosophy, and rationale

- *Number Tools,* Volumes 1 and 2, which are a series of blackline masters that serve as review sheets or practice pages involving number issues/basic skills

- *News in Numbers,* which is a set of additional activities that can be inserted between or within other units; it includes a number of measurement problems that require estimation.

Thank you for choosing *Mathematics in Context.* We wish you success and inspiration!

Sincerely,

The Mathematics in Context Development Team

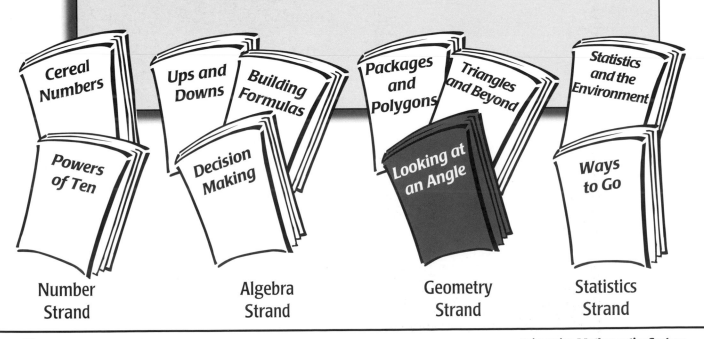

| Number Strand | Algebra Strand | Geometry Strand | Statistics Strand |

Overview

How to Use This Book

This unit is one of 40 for the middle grades. Each unit can be used independently; however, the 40 units are designed to make up a complete, connected curriculum (10 units per grade level). There is a Student Book and a Teacher Guide for each unit.

Each Teacher Guide comprises elements that assist the teacher in the presentation of concepts and in understanding the general direction of the unit and the program as a whole. Becoming familiar with this structure will make using the units easier.

Each Teacher Guide consists of six basic parts:

- Overview
- Student Materials and Teaching Notes
- Assessment Activities and Solutions
- Glossary
- Blackline Masters
- Try This! Solutions

Overview

Before beginning this unit, read the Overview in order to understand the purpose of the unit and to develop strategies for facilitating instruction. The Overview provides helpful information about the unit's focus, pacing, goals, and assessment, as well as explanations about how the unit fits with the rest of the *Mathematics in Context* curriculum.

Student Materials and Teaching Notes

This Teacher Guide contains all of the student pages (except the Try This! activities), each of which faces a page of solutions, samples of students' work, and hints and comments about how to facilitate instruction. Note: Solutions for the Try This! activities can be found at the back of this Teacher Guide.

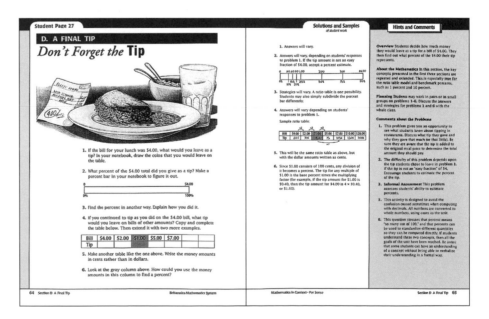

Each section within the unit begins with a two-page spread that describes the work students do, the goals of the section, new vocabulary, and materials needed, as well as providing information about the mathematics in the section and ideas for pacing, planning instruction, homework, and assessment.

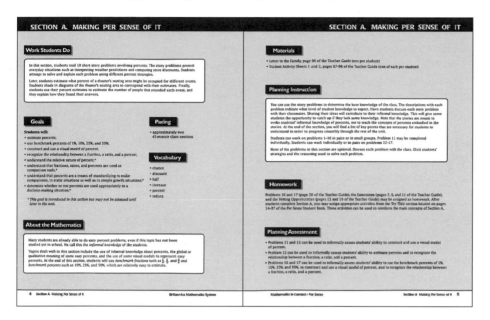

Assessment Activities and Solutions

Information about assessment can be found in several places in this Teacher Guide. General information about assessment is given in the Overview; informal assessment opportunities are identified on the teacher pages that face each student page; and the Assessment Activities section of this guide provides formal assessment opportunities.

Glossary

The Glossary defines all vocabulary words listed on the Section Opener pages. It includes the mathematical terms that may be new to students, as well as words associated with the contexts introduced in the unit. (Note: The Student Book does not have a glossary. This allows students to construct their own definitions, based on their personal experiences with the unit activities.)

Blackline Masters

At the back of this Teacher Guide are blackline masters for photocopying. The blackline masters include a letter to families (to be sent home with students before beginning the unit), several student activity sheets, and assessment masters.

Try This! Solutions

Also included in the back of this Teacher Guide are the solutions to several Try This! activities—one related to each section of the unit—that can be used to reinforce the unit's main concepts. The Try This! activities are located in the back of the Student Book.

Unit Focus

Looking at an Angle introduces students to the study of right triangles. The unit begins with an investigation of vision lines. Students explore right triangles formed by vision lines in canyons, on boats, and in cars, and by rays of light that cause shadows. Then students consider right triangles formed by ladders leaning against a wall and by the flight paths of hang gliders. Students study the relationship of the angle formed by a line and a plane and explore the ratio known as the tangent. Students discover the correspondences between lines in different contexts: the vision lines that show the area of a blind spot, the sun rays that cause parallel shadows, the light rays from lamps with radial shadows, the flight path of a hang glider, and a ladder leaning against a wall.

Mathematical Content

- investigating vision lines and vision angles
- investigating blind spots and blind areas
- investigating two- and three-dimensional objects in a plane
- exploring photographs and drawings in different projections
- investigating shadows caused by a nearby light source or caused by the sun
- measuring angles
- drawing top, side, and front views in relation to three-dimensional views
- finding steepness as measured by a ratio or by an angle
- finding tangent and glide ratios

Prior Knowledge

This unit assumes students have an understanding of the following:
- using a ratio table
- using a compass card or protractor to measure angles
- relationships between ratios, decimals, and fractions
- equivalent ratios, decimals, and fractions
- computing areas
- information that is embedded in two-dimensional representations of three-dimensional situations (top, side, and front views)
- properties of triangles (especially right triangles)
- the construction of triangles
- scale as used in drawings and maps
- line graphs

Planning and Preparation
Pacing: 16–17 days

Section	Work Students Do	Pacing*	Materials
A. Now You See It, Now You Don't	■ invent the concept of vision lines ■ understand blind spots, blind areas, and vision angles ■ draw vision lines ■ construct and measure blind areas ■ measure angles	5–6 days	■ Letter to the Family (one per student) ■ Student Activity Sheets 1–6 and 8–9 (one of each per student) ■ Student Activity Sheet 7 (several copies per group of students) ■ See page 5 of the Teacher Guide for a complete list of the materials and quantities needed.
B. Shadows and Blind Spots	■ discover the similarities between blind spots and shadows ■ discover differences between shadows caused by a nearby light source and by the sun ■ discover properties of shadows caused by the sun ■ construct shadows in two- and three-dimensional representations	4 days	■ Student Activity Sheets 10–16 (one of each per student) ■ See page 39 of the Teacher Guide for a complete list of the materials and quantities needed.
C. Shadows and Angles	■ explore steepness ■ measure steepness by finding the height-to-distance ratio and by finding the acute angle formed at the base of the right triangle	3 days	■ Student Activity Sheets 17–18 (one of each per student) ■ See page 65 of the Teacher Guide for a complete list of the materials and quantities needed.
D. Glide Angles	■ explore glide ratios ■ explore the relationship between glide ratios and glide angles ■ make relative comparisons using glide ratios ■ learn to use formal tangent notation ■ use tangents to solve problems	4 days	■ Student Activity Sheet 18 (one per student) ■ See page 79 of the Teacher Guide for a complete list of the materials and quantities needed.

* One day is approximately equivalent to one 45-minute class session.

Preparation

In the *Teacher Resource and Implementation Guide* is an extensive description of the philosophy underlying both the content and the pedagogy of the *Mathematics in Context* curriculum. Suggestions for preparation are also given in the Hints and Comments columns of this Teacher Guide. You may want to consider the following:

- Work through the unit before teaching it. If possible, take on the role of the student and discuss your strategies with other teachers.

- Use the overhead projector for student demonstrations, particularly with overhead transparencies of the student activity sheets and any manipulatives used in the unit.

- Invite students to use drawings and examples to illustrate and clarify their answers.

- Allow students to work at different levels of sophistication. Some students may need concrete materials, while others can work at a more abstract level.

- Provide opportunities and support for students to share their strategies, which often differ. This allows students to take part in class discussions and introduces them to alternative ways to think about the mathematics in the unit.

- In some cases, it may be necessary to read the problems to students or to pair students to facilitate their understanding of the printed materials.

- A list of the materials needed for this unit is in the chart on page xiii.

- Try to follow the recommended pacing chart on page xiii. You can easily spend much more time on this unit than the number of class periods indicated. Bear in mind, however, that many of the topics introduced in this unit will be revisited and covered more thoroughly in other *Mathematics in Context* units.

Resources

For Teachers	For Students
Books and Magazines • *Mathematics Assessment: Myths, Models, Good Questions, and Practical Suggestions,* edited by Jean Kerr Stenmark (Reston, Virginia: The National Council of Teachers of Mathematics, Inc., 1991)	**Software** • *Gliding* (available from Sunburst, Pleasantville, NY.)

Assessment

Planning Assessment

In keeping with the NCTM Assessment Standards, valid assessment should be based on evidence drawn from several sources. (See the full discussion of assessment philosophies in the *Teacher Resource and Implementation Guide*.) An assessment plan for this unit may draw from the following sources:

- Observations—look, listen, and record observable behavior.

- Interactive Responses—in a teacher-facilitated situation, note how students respond, clarify, revise, and extend their thinking.

- Products—look for the quality of thought evident in student projects, test answers, worksheet solutions, or writings.

These categories are not meant to be mutually exclusive. In fact, observation is a key part in assessing interactive responses and also a key to understanding the end results of projects and writings.

Ongoing Assessment Opportunities

- **Problems within Sections**
  To evaluate ongoing progress, *Mathematics in Context* identifies informal assessment opportunities and the goals that these particular problems assess throughout the Teacher Guide. There are also indications as to what you might expect from your students.

- **Section Summary Questions**
  The summary questions at the end of each section are vehicles for informal assessment (see Teacher Guide pages 36, 62, 76, and 102).

End-of-Unit Assessment Opportunities

In the back of this Teacher Guide, there are 11 assessment problems that can be completed in one or two 45-minute class periods. For a more detailed description of these assessment problems, see the Assessment Overview (Teacher Guide pages 104 and 105).

In addition, students can write their own problems involving blind spots, vision lines, shadows, steepness, height-to-distance ratios, and tangents. Students should also write the solutions, stating their assumptions and showing their calculations.

You may also wish to design your own culminating project or let students create one that will tell you what they consider important in the unit. For more assessment ideas, refer to the chart on pages xvi and xvii.

Goals and Assessment

In the *Mathematics in Context* curriculum, unit goals, categorized according to cognitive procedures, relate to the strand goals and to the NCTM Curriculum and Evaluation Standards. Additional information about these goals is found in the *Teacher Resource and Implementation Guide.* The *Mathematics in Context* curriculum is designed to help students develop their abilities so that they can perform with understanding in each of the categories listed below. It is important to note that the attainment of goals in one category is not a prerequisite to attaining those in another category. In fact, students should progress simultaneously toward several goals in different categories.

	Goal	Ongoing Assessment Opportunities	End-of-Unit Assessment Opportunities
Conceptual and Procedural Knowledge	**1.** understand the concepts of vision line, vision angle, and blind spot	**Section A** p. 34, #29 p. 36, #30	Investigating the Mammoth Rocks, p. 136, #1, #2, #3
	2. understand the concept of steepness	**Section C** p. 76, #15	
	3. understand the concept of glide ratio, or tangent	**Section D** p. 86, #6 p. 94, #16 p. 102, #30	Investigating the Mammoth Rocks, p. 138, #10, #11
	4. construct vision lines and blind spots (or light rays and shadows) in two- and three-dimensional representations	**Section A** p. 36, #30 **Section B** p. 56, #17	Investigating the Mammoth Rocks, p. 136, #3 p. 137, #6 p. 138, #8
	5. measure blind spots (or shadows)	**Section A** p. 34, #27	Investigating the Mammoth Rocks, p. 136, #4 p. 137, #5
	6. measure angles	**Section C** p. 72, #10 p. 76, #15	Investigating the Mammoth Rocks, p. 138, #9
	7. understand the difference between shadows caused by a nearby light source and shadows caused by the sun	**Section B** p. 54, #13–#15 p. 62, #26	
	8. make scale drawings of situations involving steepness	**Section C** p. 72, #10 **Section D** p. 84, #4	
	9. understand the ratio between an object and its shadow caused by the sun for different times of the day and the year	**Section B** p. 54, #13–#15 p. 56, #17 p. 58, #22	Investigating the Mammoth Rocks, p. 137, #6, #7 p. 138, #8

Goal	Ongoing Assessment Opportunities		End-of-Unit Assessment Opportunities
Reasoning, Communicating, Thinking, and Making Connections	**10.** make relative comparions involving steepness problems	**Section A** p. 34, #28 **Section C** p. 76, #15 **Section D** p. 84, #4 p. 86, #6 p. 100, #26	
	11. understand the relationship among steepness, angle, and height-to-distance ratio	**Section A** p. 20, #17c **Section B** p. 54, #13–#15 **Section C** p. 76, #15 **Section D** p. 100, #26	Investigating the Mammoth Rocks, p. 138, #11
	12. choose appropriate views (top, side, or front) to draw situations involving steepness	**Section A** p. 36, #30	Investigating the Mammoth Rocks, p. 138, #11

Goal	Ongoing Assessment Opportunities		End-of-Unit Assessment Opportunities
Modeling, Nonroutine Problem-Solving, Critically Analyzing, and Generalizing	**13.** understand the correspondences between contexts involving steepness that may be represented with a right triangle	**Section B** p. 62, #25 **Section D** p. 96, #20	
	14. use ratios to solve problems involving steepness	**Section A** p. 34, #27 **Section C** p. 76, #15 **Section D** p. 84, #4	Investigating the Mammoth Rocks, p. 138, #11
	15. solve problems involving tangents	**Section D** p. 100, #25, #27, #28, #29 p. 102, #31	Investigating the Mammoth Rocks, p. 138, #10, #11

More about Assessment

Scoring and Analyzing Assessment Responses

Students may respond to assessment questions with various levels of mathematical sophistication and elaboration. Each student's response should be considered for the mathematics that it shows, and not judged on whether or not it includes an expected response. Responses to some of the assessment questions may be viewed as either correct or incorrect, but many answers will need flexible judgment by the teacher. Descriptive judgments related to specific goals and partial credit often provide more helpful feedback than percent scores.

Openly communicate your expectations to all students, and report achievement and progress for each student relative to those expectations. When scoring students' responses, try to think about how they are progressing toward the goals of the unit and the strand.

Student Portfolios

Generally, a portfolio is a collection of student-selected pieces that is representative of a student's work. A portfolio may include evaluative comments by you or by the student. See the *Teacher Resource and Implementation Guide* for more ideas on portfolio focus and use.

A comprehensive discussion about the contents, management, and evaluation of portfolios can be found in *Mathematics Assessment: Myths, Models, Good Questions, and Practical Suggestions*, pp. 35–48.

Student Self-Evaluation

Self-evaluation encourages students to reflect on their progress in learning mathematical concepts, their developing abilities to use mathematics, and their dispositions toward mathematics. The following examples illustrate ways to incorporate student self-evaluations as one component of your assessment plan.

- Ask students to comment, in writing, on each piece they have chosen for their portfolios and on the progress they see in the pieces overall.

- Give a writing assignment entitled "What I Know Now about [a math concept] and What I Think about It." This will give you information about each student's disposition toward mathematics as well as his or her knowledge.

- Interview individuals or small groups to elicit what they have learned, what they think is important, and why.

Suggestions for self-inventories can be found in *Mathematics Assessment: Myths, Models, Good Questions, and Practical Suggestions*, pp. 55–58.

Summary Discussion

Discuss specific lessons and activities in the unit—what the student learned from them and what the activities have in common. This can be done in whole-class discussion, in small groups, or in personal interviews.

Connections across the *Mathematics in Context* Curriculum

Looking at an Angle is the seventh unit in the geometry strand. The map below shows the complete *Mathematics in Context* curriculum for grade 7/8. This indicates where the unit fits in the geometry strand and in the overall picture.

A detailed description of the units, the strands, and the connections in the *Mathematics in Context* curriculum can be found in the *Teacher Resource and Implementation Guide*.

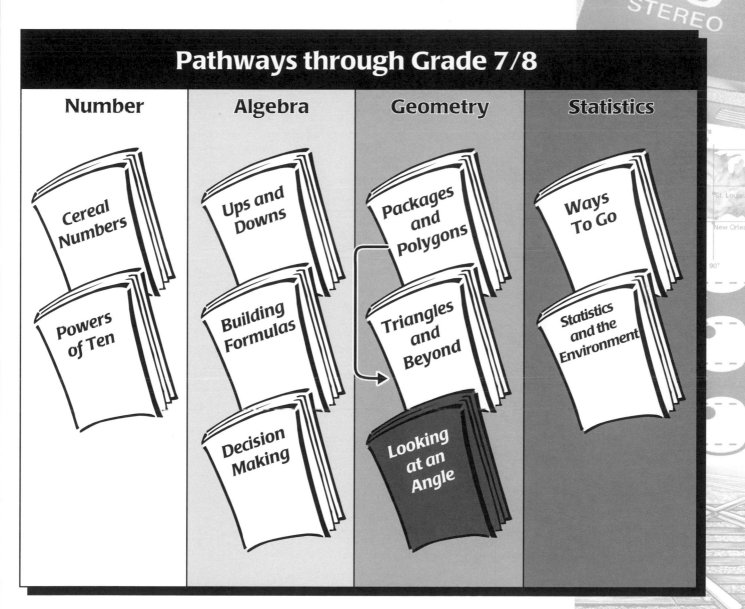

Pathways through Grade 7/8

Number	Algebra	Geometry	Statistics
Cereal Numbers	Ups and Downs	Packages and Polygons	Ways To Go
Powers of Ten	Building Formulas	Triangles and Beyond	Statistics and the Environment
	Decision Making	Looking at an Angle	

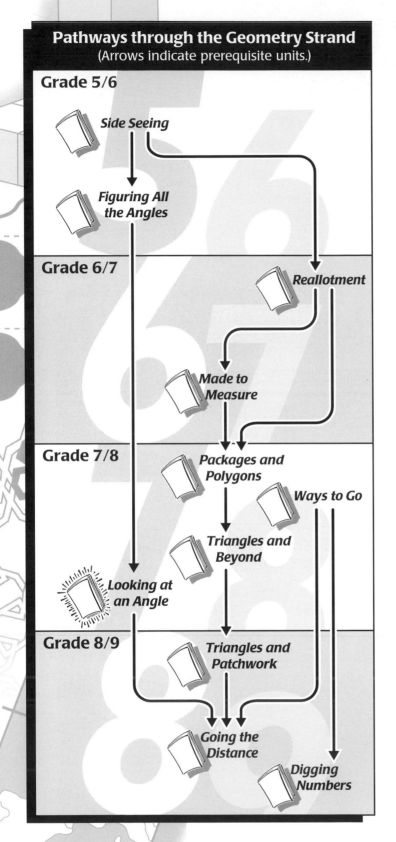

Pathways through the Geometry Strand
(Arrows indicate prerequisite units.)

Grade 5/6

Side Seeing

Figuring All the Angles

Grade 6/7

Reallotment

Made to Measure

Grade 7/8

Packages and Polygons

Ways to Go

Triangles and Beyond

Looking at an Angle

Grade 8/9

Triangles and Patchwork

Going the Distance

Digging Numbers

Connections within the Geometry Strand

On the left is a map of the geometry strand; this unit, *Looking at an Angle,* is highlighted.

The major themes in the geometry strand are orientation and navigation, shape and construction, and visualization and representation. *Looking at an Angle,* the seventh unit in the geometry strand, builds on knowledge students acquired in previous units. For example, in the unit *Figuring All the Angles,* students learned how to measure angles. In the unit *Side Seeing,* students were introduced to different views of objects. Students began computing areas when they worked on the unit *Reallotment.* The unit *Triangles and Beyond* introduced students to properties of triangles.

The focus of *Looking at an Angle* is on different contexts that may be represented with a right triangle. In each context there is a height and a distance. The height may be that of an object blocking one's view, an object casting a shadow, the point where a ladder touches a wall, or the height of a cliff from which a hang glider is to be launched. The distance may be the length of a blind area, the length of a shadow, the distance between the foot of a ladder and the wall, or the ground distance covered by a hang glider as it flies. The unit deals with the ratio of height to distance: the steepness, glide ratio, or tangent. Students investigate the relationship between the angle between a right triangle's base and its hypotenuse, and its tangent.

The grade 8/9 unit *Going the Distance* builds on the knowledge students acquire in this unit.

The Geometry Strand

Grade 5/6

Side Seeing
Exploring the relationship between three-dimensional shapes and drawings of them, seeing from different points of view, and building structures from drawings.

Figuring All the Angles
Estimating and measuring angles and investigating direction, vectors, and rectangular and polar coordinates.

Grade 7/8

Packages and Polygons
Recognizing geometric shapes in real objects and representations, constructing models, and investigating properties of regular and semi-regular polyhedra.

Looking at an Angle
Recognizing vision lines in two and three dimensions; identifying and drawing shadows and blind spots; identifying the isomorphism of vision lines, light rays, flight paths, and so forth; understanding the relationship between angles and the tangent ratio; and computing with the tangent ratio.

Ways to Go
Reading and interpreting different kinds of maps, comparing different types of distances, progressing from one two-dimensional model to another (from a diagram to a map to a photograph to a graph), and drawing graphs and networks. (*Ways to Go* is also in the statistics strand.)

Triangles and Beyond
Exploring the interrelationships of the sides and angles of triangles as well as the properties of parallel lines and quadrilaterals, constructing triangles, and using transformations to become familiar with the concepts of congruence and similarity.

Grade 6/7

Reallotment
Measuring regular and irregular areas; discovering links between area, perimeter, surface area, and volume; and using English and metric units.

Made to Measure
Measuring length (including circumference), volume, and surface area using metric units.

Grade 8/9

Triangles and Patchwork
Understanding similarity and using it to find unknown measurements for similar triangles and developing the concept of ratio through tessellation.

Going the Distance
Using the Pythagorean theorem to investigate distances, scales, and vectors and using slope, tangent, area, square root, and contour lines.

Digging Numbers
Using the properties of height, diameter, and radius to determine whether or not various irregular shapes are similar; predicting length using graphs and formulas; exploring the relationship between three-dimensional shapes and drawings of them; and using length-to-width ratios to classify various objects. (*Digging Numbers* is also in the statistics strand.)

Connections with Other *Mathematics in Context* Units

The following mathematical topics that are included in the unit *Looking at an Angle* are introduced or further developed in other *Mathematics in Context* units.

Prerequisite Topics

Topic	Unit	Grade
angles	*Figuring All the Angles*	5/6
	Made to Measure	6/7
ratios, fractions, decimals	*Fraction Times**	6/7
	*More or Less**	6/7
	*Ratios and Rates**	6/7
area	*Reallotment*	6/7
two-dimensional representations	*Side Seeing*	5/6
triangles	*Triangles and Beyond*	7/8
scale	*Grasping Sizes**	5/6
	Figuring All the Angles	5/6
line graphs	*Tracking Graphs***	6/7
maps	*Figuring All the Angles*	5/6
ratio tables	*Grasping Sizes**	5/6
	*Per Sense**	5/6
	*Number Tools**	

Topics Revisited in Other Units

Topic	Unit	Grade
angles	*Packages and Polygons*	7/8
	Triangles and Beyond	7/8
	Triangles and Patchwork	8/9
	Going the Distance	8/9
fair exchange	*Decision Making***	7/8
slope	*Graphing Equations***	8/9
	Going the Distance	8/9
triangles	*Triangles and Patchwork*	8/9
maps	*Going the Distance*	8/9
tangent	*Going the Distance*	8/9

* These units in the number strand also help students make connections to ideas about geometry.
** These units in the algebra strand also help students make connections to ideas about geometry.

Student
Materials
and Teaching
Notes

Student Book
Table of Contents

Dear Student,

Welcome to *Looking at an Angle*.

In this unit you will learn about vision lines and blind areas. Have you ever been on one of the top floors of a tall office or apartment building? When you looked out the window, were you able to see the sidewalk directly below the building? If you could see the sidewalk, it was in your field of vision; if you could not see the sidewalk, it was in your blind spot.

The relationship between vision lines and light rays and the relationship between blind spots and shadows are some of the topics that you will explore in this unit. Have you ever noticed how the length of a shadow varies according to the time of day? As part of an activity, you will measure the length of the shadow of a stick and the corresponding angle of the sun at different times of the day. You will then determine how the angle of the sun affects the length of a shadow.

Besides looking at the angle of the sun, you will also study the angle that a ladder makes with the floor when it is leaning against a wall and the angle that a descending hang glider makes with the ground. You will learn two different ways to identify the steepness of an object: the angle the object makes with the ground and the tangent of that angle.

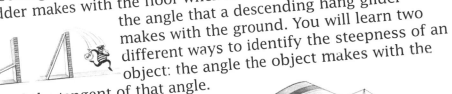

We hope you enjoy discovering the many ways of "looking at an angle."

Sincerely,

The Mathematics in Context Development Team

Work Students Do

In this section, students investigate blind spots. They begin at the Grand Canyon, deciding what conditions are necessary for a hiker at the rim to be able to see the Colorado River below. Students build models of canyons and construct vision lines to help them visualize the situation. They then investigate how the size and shape of a ship affect the captain's vision lines as well as the size, location, and shape of the captain's blind spot. Students measure the angle between the captain's vision line and the water. They outline the blind spots of toy boats on grid paper. They also build tugboats with cubes, find vision lines, and delineate blind areas. Students then investigate the blind spots of cars. Finally, students describe situations in their daily lives that involve blind spots.

Goals

Students will:

- understand the concepts of vision line, vision angle, and blind spot;
- understand the concept of steepness;*
- construct vision lines and blind spots (or light rays and shadows) in two- and three-dimensional representations;
- measure blind spots (or shadows);
- measure angles;*
- make relative comparisons involving steepness problems;
- understand the relationship among steepness, angle, and height-to-distance ratio;
- choose appropriate views (top, side, or front) to draw situations involving steepness;
- use ratios to solve problems involving steepness.

  * These goals are introduced in this section and are assessed in other sections of the unit.

Pacing

- approximately five or six 45-minute class sessions

Vocabulary

- blind area
- blind spot
- vision line

About the Mathematics

This section requires that students visualize situations in three dimensions and represent these situations on paper. These are important geometric skills. The context used here is blind spots. The captain of a boat has a blind spot that is caused by the deck of the boat. In exploring a captain's vision line (running from the captain's eyes past the edge of the boat's deck and continuing in a straight line to the water) and how it relates to the size of his or her blind spot, students begin to understand the concept of slope. Slope is the ratio of change in height over change in length. Students measure the angles formed by various vision lines and the water. The smaller the angle that a vision line makes with the water, the larger the blind spot will be. Conversely, the steeper the vision line, the larger the angle of elevation and the smaller the blind spot.

Materials

- Letter to the Family, page 116 of the Teacher Guide (one per student)
- Student Activity Sheets 1–6 and 8 and 9, pages 117–122, 124, and 125 of the Teacher Guide (one of each per student)
- Student Activity Sheet 7, page 123 of the Teacher Guide (several copies per group of students)
- See the Hints and Comments columns on each right-hand page of Section A for a complete list of the materials and quantities needed.

Planning Instruction

To begin this section, you might ask students whether they have ever been to the Grand Canyon. If they have, ask how close they went to the edge, and whether they saw the Colorado River at the bottom of the gorge. Many landforms besides the Grand Canyon create blind spots. You may also wish to ask students about their experiences with mountains, hills, and valleys. If so, discuss the blind spots that are created by these landforms.

Students may work on problems 4–6 in small groups or as a class. Problems 7, 19, 27–29, and 30 may be done individually. Students may work on problem 21 individually or in small groups. They may work on problems 25 and 26 in pairs. Students may do the remaining problems in small groups.

Problems 20, 21, 27, 28, and 29, and the activities on pages 12 and 16, are optional. If time is a concern, you may omit these problems or assign them as homework.

Homework

Problems 19 (page 22 of the Teacher Guide), 20 (page 24 of the Teacher Guide), 21 (page 26 of the Teacher Guide), and 27–29 (page 34 of the Teacher Guide) and the activities on pages 12 and 16 of the Teacher Guide may be assigned for homework. The Extension (page 33 of the Teacher Guide), the Writing Opportunities (pages 11, 17, and 37 of the Teacher Guide), and the Bringing Math Home activities (pages 35 and 37 of the Teacher Guide) may also be assigned as homework. After students complete Section A, you may assign appropriate activities from the Try This! section, located on pages 47–50 of the *Looking at an Angle* Student Book. The Try This! activities reinforce the key mathematical concepts introduced in this section.

Planning Assessment

- Problem 17c can be used to informally assess students' ability to understand the relationship among steepness, angle, and height-to-distance ratio.
- Problem 27 can be used to informally assess students' ability to measure blind spots (or shadows) and use ratios to solve problems involving steepness.
- Problem 28 can be used to informally assess students' ability to make relative comparisons involving steepness problems.
- Problem 29 can be used to informally assess students' ability to understand the concept of a vision angle.
- Problem 30 can be used to informally assess students' ability to understand the concepts of vision line, vision angle, and blind spot; construct vision lines and blind spots in two- and three-dimensional representations; and choose appropriate views (top, side, or front) to draw situations involving steepness.

A. NOW YOU SEE IT, NOW YOU DON'T

The Grand Canyon

The Grand Canyon is one of the most famous natural wonders of the world. It is a huge gorge, cut by the Colorado River into the high plateau of northwestern Arizona. The total length of the gorge is 446 kilometers. Approximately 90 kilometers of the gorge are located in Grand Canyon National Park. The north rim of the canyon (the Kaibab Plateau) is about 2,500 meters above sea level.

The above photo shows part of the Colorado River. You can see the river on the left side of the photo.

1. Why can't you see the continuation of the river on the lower right side of the photo?

1. Explanations will vary, but students should indicate that something is blocking their view. Sample student explanations:

   The ledge in front of us is hiding the Colorado River.

   The river is a long way down and the rocks are in the way.

Overview Students look at a photograph of the Grand Canyon. They determine why the Colorado River cannot be seen in the lower right portion of the photograph.

About the Mathematics An important mathematical goal for this section is to have students understand the concept of the *vision line*, an imaginary straight line from your eye to an object. Do not, however, mention this idea yet. At this point, it is enough for them to realize that something may be hidden from view because something is in the way. Later in this section, the concept of a "blind spot" will be introduced.

Planning Students may work on problem **1** in small groups. Discuss students' answers with the class.

Comments about the Problems

1. Encourage students to focus on the location of the photographer's camera, the location of the rim of the canyon, and the visibility of the river. Most students know from playing hide-and-seek that objects or persons may be hidden from view when something is in the way. If students have difficulty, you could set up a simple classroom situation to make this point clear. For example, if someone is approaching the classroom, some students may be able to see the approaching person through the open door. However, other students may find that their view of the approaching person is blocked by a wall.

In fact, the Colorado River can barely be seen from most viewpoints in Grand Canyon National Park.

The picture on the right shows a hiker on the north rim overlooking a portion of the canyon.

2. Can the hiker see the river directly below her? Explain.

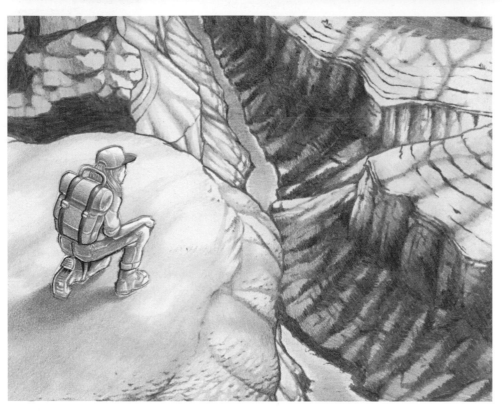

Shown below are a photograph and a drawing of the same area of the Grand Canyon. From the drawing, you can see that the canyon walls are shaped like stairs.

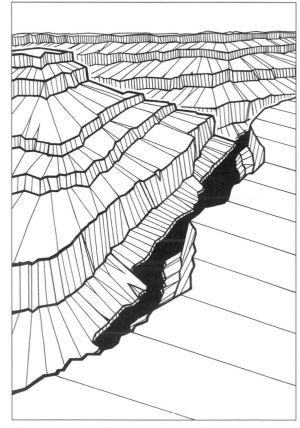

3. What other differences do you notice between the photo and the drawing?

2. Answers will vary. Sample response:

The hiker is kneeling on a rock that is blocking her view. She can probably see the river only if she looks to the left or the right.

3. Answers will vary. Sample response:

The drawing simplifies the canyon. Details that can be seen on the photograph, such as roads, trees, differences in the color of the rock formations, and shadows are lost in the drawing. Because these details are left out of the drawing, other features of the canyon can be seen more clearly. It is easier to see the river. It is also easier to see that the rock formations are shaped like stairs.

Overview Students visualize how much of the Colorado River a hiker on the rim can see. Students also consider the difference between a drawing and a photograph of the Grand Canyon.

About the Mathematics In problem **1,** students were asked to work with information supplied by a photograph. Problem **2** is different in the sense that students have to put themselves in the position of the hiker to imagine this person's view of the river. Problem **2** is aiming at the idea of vision lines. Students may actually come up with the idea of using string or something else to show such a line.

Planning Students may work on problems **2** and **3** in small groups. You may want to discuss students' answers with the whole class.

Comments about the Problems

2. Some students may indicate with gestures what areas of the canyon the hiker will probably see.

3. Some students may find the drawing easier to understand, while others may prefer the photograph. In either case, students should realize that the drawing simplifies the canyon's rock formations.

Activity

The Table Canyon Model

In this activity, you will build your own "table canyon" to investigate how much of the "river" can be seen from different perspectives.

   i. Place two tables parallel to each other, with enough room between them for another table to fit.

  ii. Hang large sheets of paper from the tables to the floor as shown in the drawing on the right. (The paper represents the canyon walls, and the floor between the two tables represents the river.)

 iii. Sit behind one of the tables, and have another student sit behind the other.

 iv. Have a third student make at least three marks on each canyon wall to show the lowest place visible to each of you.

4. a. Can either of you see the river below? Explain.

   b. On which wall are the marks the highest? Explain.

   c. Are all the marks on one wall the same height? Explain.

5. Describe some possible changes that would allow you to see the river. How does each change affect what you can see?

6. a. If there were a boat on the river, where could the boat be located so that both of you can see it?

   b. What would happen if the boat moves closer to one of the canyon walls?

7. Write a report on this activity describing your investigations and discoveries. You may want to use the terms *visible, not visible,* and *blind spot* in your report.

4. a. Answers will vary. Students may or may not be able to see the rivers, depending on how they are seated and their height.

b. Answers will vary. Students will find that the marks of a taller person or a person leaning forward are lower.

c. Yes. Explanations will vary. Sample explanation:

The marks on the opposite wall should all be at the same height if the tables are parallel. The distance down the canyon that each student can see is determined by the angle of the student's eyes relative to the first obstruction that blocks his or her view.

5. Answers will vary. Sample student response:

There are several things I can do to see the river. I can sit up straight, stand up, lean forward, or widen the canyon by moving one or both tables.

6. Answers will vary. Sample student responses:

a. If we could both see just enough of the river to see the middle, then we could see a boat coming down the middle of the river.

b. If the boat moved closer to one of the canyon walls, then whoever was sitting next to that wall would no longer be able to see the boat.

7. Reports will vary. Sample student report:

At first, the river was not visible at all. Then we moved the tables farther apart so that the other side of the river was visible. When the boat came down my side of the river, it was in my blind spot and it didn't matter how far up or down the river the boat was. The boat only became visible to me when it moved to the other side of the river.

Materials 3 ft × 5 ft length of paper (two sheets per group of students); markers (one per group of students); string (one roll per group of students); scissors (one pair per student or group of students); toy boats with flat bottoms, optional (one per group of students)

Overview Students build their own canyons out of tables and paper. They investigate the boundaries of what they can and cannot see.

About the Mathematics Students should discover the idea of *vision lines* in answering the problems on Student Book pages 1–3. The eye is used as the starting point of the vision line while the dot placed on the opposite canyon wall is the ending point. Informally, students investigate *angle of elevation, steepness,* and *blind spots.*

Planning When students construct their model canyons, make sure that the sheets of paper representing the canyon walls are hanging straight down to the floor. Students may work on problems **4–6** in small groups or as a class. They may work on problem **7** individually. Make sure that every student gets to sit in a chair and investigate vision lines. At the end of the activity, have students discuss their results with the whole class.

Comments about the Problems

4. c. This problem can lead to an interesting discussion because the results may not be what students expected. Some students may have expected the mark in the middle to be lower than those on the sides, thinking that the "shortest route" would allow them to see further down the wall. If students have difficulty, you might have them represent the line from each student's eyes to each of the three marks with string.

6. You may wish to have students experiment using a toy boat. Students may put the boat in several different places. For example, they could put it in a visible area and move it closer and closer toward a canyon wall until it disappears from view.

Writing Opportunity You may ask students to write their answers to problem **7** in their journals. Writing a report on this activity encourages students to reflect on what they did, and gives you something to refer to during the rest of the unit.

Activity

The Paper Canyon Model

In this activity, you will build a simplified paper model of part of a canyon, as shown on the right. This model can help you to explain why it is sometimes difficult to see the river from different locations on the rim of a canyon.

i. Cut out the nets on **Student Activity Sheets 1** and **2.**

ii. Fold the nets along the dotted lines and then tape the sides of each net to make two canyon walls.

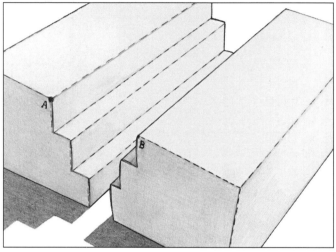

iii. Place the canyon walls so that they face each other as shown in the picture on the right.

With this model, you can construct vision lines with a thin, straight object, such

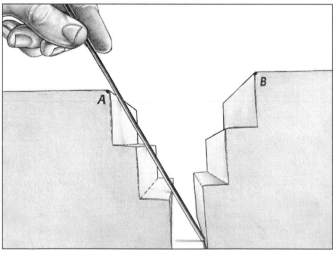

as a straw. *Vision lines* are imaginary, straight lines from a person's eyes to an object that show what is in the person's field of vision.

Students' model canyons should look similar to those shown on page 4 of the Student Book.

Materials Student Activity Sheets 1 and 2 (one of each per student); scissors (one pair per student); $8\frac{1}{2}$" × 11" construction paper (two sheets per group); tape (one dispenser per group); straws (one per group)

Overview Students build paper models of canyons. They use straws to model vision lines.

About the Mathematics Student Book page 4 gives a definition of *vision line*. By looking at different objects, you create many vision lines from your eye to all of the objects. In the context of the Grand Canyon, we are interested in the specific lines that show us the boundary of what we can or cannot see. Students should understand this concept by working with the canyon model and a straw; they should realize that the straw has to go from the top of the canyon wall down toward the river, touching the ledges. Discuss this with your students.

Planning Students may work on this activity in small groups. The activity is optional. If time is a concern, it may be omitted or assigned as homework. If you decide to omit the activity, you might want to make a model for the class to look at. You may want students to trace the diagrams on Student Activity Sheets 1 and 2 onto construction paper, which is thicker and will make a sturdier canyon model.

The following drawing shows a side view of the paper canyon model that you built on page 4.

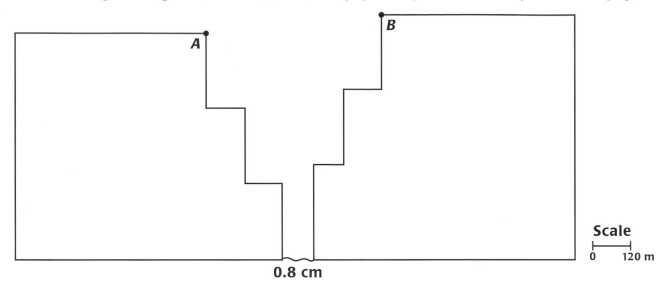

0.8 cm

Scale
0   120 m

To answer problems **8–12,** use either the model you made in the previous activity and a straw or use **Student Activity Sheet 3** and a straightedge. If you use your paper model, be sure to place the canyon walls as shown above, leaving a space of 0.8 centimeter for the river.

8. Is it possible to see the river from point *A* on the left rim? Why or why not?

9. What is the actual height of the left canyon wall represented by the above scale model?

10. What is the actual width of the river represented by the above scale model?

11. If the river were 1.2 centimeters wide in the above scale model, could it be seen from point *A*? Explain.

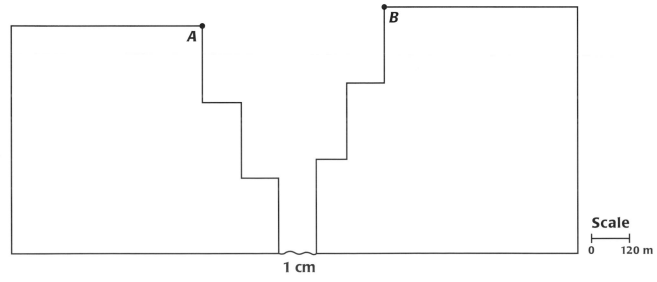

1 cm

Scale
0   120 m

12. In the above drawing, the river is 1 centimeter wide. Is it possible to see the river from point *B*? If not, which ledge is blocking your view? Explain.

Solutions and Samples
of student work

Hints and Comments

8. No. Explanations will vary. Sample explanation:

It is not possible to see the river from point *A*, because the straw (the vision line) overshoots the river and hits the opposite canyon wall, as shown below.

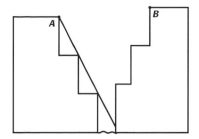

9. The height is 720 meters. Point *A* is six centimeters above the river in the model. Every centimeter in the drawing represents 120 meters. Students may use a ratio table:

× 6

Drawing	1 cm	6 cm
Reality	120 m	720 m

10. The river is 96 meters wide. Students may use a variety of strategies to find their answers. Some students may subtract 0.2 (or $\frac{1}{5}$) of 120 meters from 120 meters; 120 meters − 24 meters = 96 meters.

11. Yes. A line of sight to at least part of the river is now clear. Students may check by drawing a new vision line, as shown below.

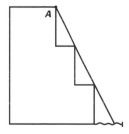

12. No. The first ledge below point *B* blocks the line of sight, as shown below.

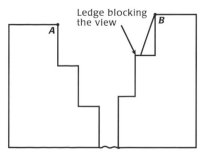

Ledge blocking the view

Materials Student Activity Sheet 3 (one per student); straws (one per group); calculators, optional (one per student); straightedges (one per student); canyon models made by students on page 4 of the Student Book, optional (one per group)

Overview Students work on problems involving scale and ratio. They refer to their paper models of the canyon and side-view drawings.

About the Mathematics Students should be familiar with drawing and interpreting side views from their work in the unit *Side Seeing.* They have to realize that a vision line starts from point *A* and extends straight until it meets with something opaque.

Students should also have a basic understanding of scale and ratio from the unit *Grasping Sizes.* They should be familiar with the use of a ratio table from the unit *Some of the Parts* or sections of *Number Tools.*

Planning Students may work on problems **8–11** in small groups. You may want to discuss problems **8** and **10** in class.

Comments about the Problems

8–12. Encourage students to use a ratio table to solve the scale problems.

8. Students may use the model or Student Activity Sheet 3. Some students may realize that there is a certain pattern to the way the left wall is constructed: two centimeters down, one centimeter over, two centimeters down, and so on. This means that the river can be seen from the rim only if it is at least one centimeter wide.

9–10. Suggest that students use a ratio table. You may also wish to have them use a calculator.

11. Students can adjust the existing drawing or make a new one.

12. Students may check their answers by drawing a vision line from point *B* to the river.

Most of the ledges of the Grand Canyon are rather wide.

13. Explain why wide canyon ledges make it difficult to see the river.

Activity

In the following activity, you will create a canyon model with a curve by building extensions for the model you made on page 4.

   i. Cut out the nets on **Student Activity Sheets 4** and **5.** Fold the nets along the dotted lines and then tape the sides together to create two canyon walls.

   ii. Place extension A next to the canyon wall on the left that is labeled with point *A*. Place extension B next to the canyon wall on the right that is labeled with point *B*.

In this expanded model, there are some good places to view the river.

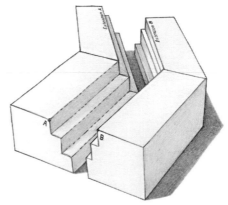

14. Make a sketch of the model shown above. On your sketch, indicate the best places on the canyon rim for viewing the river. Use arrows to show the direction in which the river is visible.

13. Explanations will vary. Sample explanations:

The wider the ledges of the canyon, the less steep, or more horizontal, the vision line becomes. This makes the blind spot larger.

Some students may illustrate their answers with a drawing:

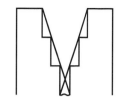

Narrow Ledges

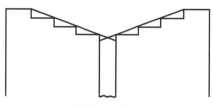

Wide Ledges

14. Drawings will vary. Sample drawing:

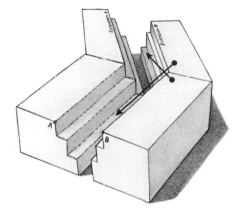

At points near the curve on the right-hand side of the canyon (wall B), an observer can see both sections of the river.

Materials Student Activity Sheets 4 and 5 (one of each per student); scissors (one pair per student); tape (one dispenser per group); straws (one per group); straightedges (one per student)

Overview Students consider the relationship between the widths of canyon ledges and the size of the blind spot for a viewer standing at the rim. They extend their paper models of the canyons to include angles and find locations on the rim where hikers would be able to see the river.

About the Mathematics *Steepness* is an important concept throughout this unit. Students do not have to use this term yet, but should develop an intuitive understanding of it.

Planning Students may work on problems **13** and **14** in small groups. The canyon-extending activity is optional. If time is a concern, you might omit this activity or assign it as homework. If you decide to omit the activity, you may want to make a model for the class to look at.

Comments about the Problems

13. Students can make drawings to illustrate their answers.

14. Students may create a top-view drawing or a perspective drawing similar to the one on page 6 of the Student Book. They can use their models and straws to check their answers. If students have difficulty, you may want to simulate the activity with the canyon tables. Four tables are needed to create the angle shown on Student Book page 6. Since there is an angle in the canyon, students may look to the left or the right to see the river. The situation is similar to the one depicted in the photograph on page 1 of the Student Book.

Writing Opportunity You may ask students to write their answers to problem **13** in their journals.

SHIPS AHOY

Picture yourself in a small rowboat rowing toward a ship that is tied to a dock. In the first picture, the captain at the helm of the ship is able to see you. As you get closer, at some point the captain is no longer able to see you.

15. Explain why the captain probably cannot see you in the last picture at the bottom of the page.

Closer

Closer

Closer

15. Explanations will vary. Sample explanation:

As you row closer to the barge, you move into the captain's blind spot. From the captain's perspective, the bow of the boat is in the way.

Overview Students investigate the blind spot of someone standing on the bridge of a ship.

About the Mathematics Again we are dealing with a person looking at something with his or her view partially obstructed by an object (in this case the bow of the boat). The students will draw vision lines, measure the angle the vision line makes with the water, and draw conclusions about the captain's blind spot.

Later in this section, *blind spot* will be referred to as the "blind area."

Planning Before students begin problem **15,** you may want to introduce some nautical terms, such as *bow, stern, bridge, crew,* and *sail.* See the Did You Know? below for more information. Students may work on problem **15** in small groups.

Comments about the Problems

15. If students have difficulty, you might remind them of the canyon table activity, where they moved a boat into a blind spot on the water by bringing it closer to a canyon wall.

Extension Ask students, *Is it true that if you cannot see the captain, the captain cannot see you?* [Basically, this statement is true. However, it is possible that you are able to see the captain's head sticking out but you cannot see the captain's eyes. In that case, the captain cannot see your eyes. So, the formulation must be more exact: If you cannot see the captain's eyes, the captain cannot see your eyes. For this reason, in traffic, people often try to make eye contact with other drivers to make sure that the other drivers see them.]

Did You Know? A ship has two main sections—the front, or the bow, and the back, or the stern. A boat is navigated from a raised platform called the bridge. Some boats have a sail, which is a piece of fabric that, with the wind, propels the ship through water. The people who work on a ship are the crew.

The shape of the ship and the captain's height and position in the ship determine what the captain can and cannot see in front of the ship. To find the captain's field of vision, you can draw a vision line (a line that extends from the captain's eyes, over the edge of the ship, and to the water).

16. For each ship shown on **Student Activity Sheet 6,** draw a vision line from the captain, over the front edge of the ship, to the water. Measure the angle between the vision line and the water. (The captain is located at the star symbol.)

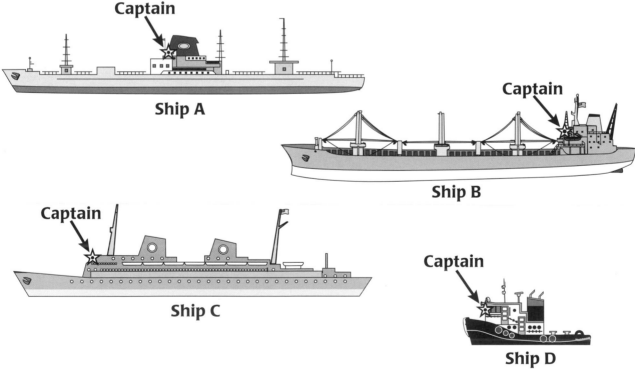

17. Compare the ships on **Student Activity Sheet 6.**

    a. On which ship is the captain's blind area the smallest? Explain.

    b. How does the shape of the ship affect the captain's view?

    c. How does the angle between the vision line and the water affect the captain's view?

16.

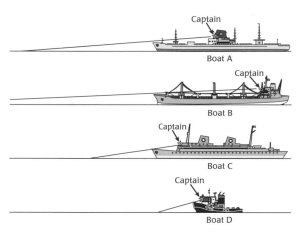

Captain — Boat A

Captain — Boat B

Captain — Boat C

Captain — Boat D

The angles of the vision lines are approximately as follows:

Boat A: 6°

Boat B: 3°

Boat C: 10°

Boat D: 23°

17. a. The captain of boat D has the smallest blind area, and the best view.

b. Explanations will vary. Sample explanation:

On boat D, the captain's bridge is close to the bow and very high. The captain's vision line is steeper than those of the captains on the other boats.

On boat B, the captain's bridge is at the back of the boat. The vision line extends a long way and is more horizontal than those of the captains on the other boats. The captain of boat B has the biggest blind spot and the worst visibility.

c. Explanations will vary. Sample explanation:

On boat D, the angle between the vision line and the water is large, and the blind spot is small. On boat B, the angle between the vision line and the water is very small, and the blind spot is large. When the angle is large there is a small blind spot and good visibility, and when the angle is small, there is a large blind spot and poor visibility.

Materials Student Activity Sheet 6 (one per student); straightedges (one per student); compass cards or protractors (one per student)

Overview Students draw vision lines and measure the angle between the vision line and the water. They draw conclusions about the size of the captain's field of vision.

Planning Students may work on problems **16** and **17** in small groups. Problem **17c** can also be used as an informal assessment. Discuss students' answers in class. You might review how to measure angles with a compass card, which is introduced in the grade 5/6 unit *Figuring All the Angles.*

Comments about the Problems

16. Make sure students draw the vision lines correctly. If students reason that there are other obstacles on the boat that are in the way, such as towers, tell them that the captain can see around those.

Students' angle measurements may vary, due to errors in measurement. Accept responses that are close to the given angle measurements in the solutions column.

17. c. Informal Assessment This problem assesses students' understanding of the relationship between the angle and the height-to-distance ratio.

Vision lines, such as the ones you drew on **Student Activity Sheet 6,** do not show everything that captains can and cannot see. For example, some ships' bridges, the area from which the captain navigates the ship, are specially constructed to improve the captain's view. The captain can walk across the bridge, from one side of the boat to the other side, to increase his or her field of vision.

On the left is a picture of the *Queen Elizabeth II.* Notice how the bridge, located between the arrows, has wings that project out on each side of the ship.

18. Explain how the wings of the bridge give the captain a better view of the water in front of the ship.

Hydrofoils have fins that raise the boat out of the water when it travels at high speeds.

19. Make two side-view drawings of a hydrofoil: one of the hydrofoil in the water traveling at a slow speed and one of it raised out of the water traveling at a high speed. Use vision lines to show the difference between the captain's view in each drawing. (You may design your own hydrofoil.)

18. The captain can walk to the ends of the wings and increase the area he or she can see directly in front and on the sides of the ship. Students may make a drawing showing how the blind spot moves as the captain walks from one side of the bridge to the other, as shown below.

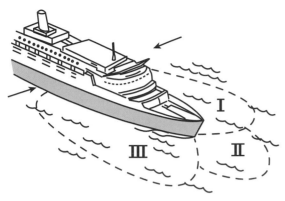

Area I indicates the blind spot when standing on the left side of the bridge.

Area II indicates the blind spot when standing in the middle of the bridge.

Area III indicates the blind spot when standing on the right side of the bridge.

19. Drawings will vary. Sample drawings:

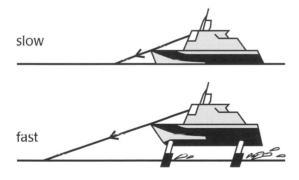

slow

fast

The higher the boat, the bigger the captain's blind spot is.

Materials straightedges (one per student); blank transparencies, optional (two per class); overhead projector, optional (one per class)

Overview Students study different boats: a boat with a wing-like bridge to improve the captain's view and a hydrofoil that is lifted from the water at high speeds.

About the Mathematics Students imagine themselves on the bridge of the boat to discover the area the captain cannot see in front of the boat. Sketching the situation can be helpful. Students may first think that their view will improve as the boat lifts out of the water, since this was the case with the previous boats: a boat with a tall bridge gives the captain a better view. However, it is not just the captain that is lifted, but the whole boat, causing the bow to block even more of the area in front of the boat.

Planning Students may work on problem **18** in small groups. They may do problem **19** individually. This problem may also be assigned as homework. Discuss students' answers in class.

Comments about the Problems

19. Homework This problem may be assigned as homework. If students have difficulty, you might prepare two transparencies: one showing a hydrofoil and a vision line and one showing the water level. Moving the transparency with the hydrofoil up from the water will help students see what happens: the vision line stays the same, the angle between the vision line and the water stays the same, but the blind spot increases.

Photo 1

Photo 2

The above photos show what a passenger in a boat can see at different speeds. In photo 1, the boat is moving slowly. In photo 2, the boat is moving more quickly.

20. Make side-view drawings of the boat for each photo. Be sure to draw vision lines and indicate the area of the water that the passenger cannot see. Explain the differences in the two views.

20. Drawings will vary. Sample drawings:

slow

blind spot

fast

blind spot

When the boat is going slowly, it is more horizontal on the water. When it is going at full speed, the front of the boat rises out of the water. As a result, the passenger in the boat can see less of the area in front of the boat.

Materials string, optional (one roll per group); straightedges (one per student); toy boats with flat bottoms, optional (one per group)

Overview Students draw vision lines and blind spots in side views of a boat going at different speeds.

Planning Students may work on problem **20** in small groups. This problem is optional. If time is a concern, it may be omitted or assigned as homework.

Comments about the Problems

20. Homework This problem may be assigned as homework. Students should indicate how and why the blind spot changes. If students have difficulty, you may want to have them model this problem using a toy boat. They could model the vision line with a piece of string.

Suppose that you are swimming in the water and a large boat is coming toward you. If you are too close to the boat, the captain may not be able to see you! In order to see a larger area of the water, a captain may travel in a zigzag course, as shown in the picture below.

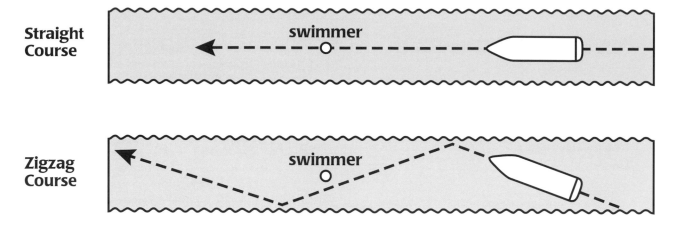

Straight Course

swimmer

Zigzag Course

swimmer

21. Explain why the captain has a better chance of seeing something in front of the boat by traveling in a zigzag course.

21. Explanations may vary. Sample response:

By zigzagging, the captain is no longer looking over the bow of the boat, but over the side. The side of the ship is closer to the bridge, so the captain's vision line over the side of the ship is steeper and provides a better view than the vision line over the bow. The drawings below show why:

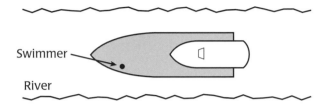

If the captain sails straight ahead, the swimmer is in the boat's blind spot.

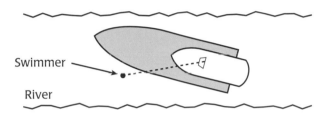

If the captain turns the boat, the swimmer is out of the blind spot.

Materials string, optional (one roll per group); toy boats with flat bottoms, optional (one per group)

Overview Students investigate how sailing a zigzag course changes the captain's field of vision.

Planning Students may work on problem **21** individually or in small groups. This problem is optional. If time is a concern, it may be omitted or assigned as homework. If you assign problem **21** as homework, be sure to discuss students' answers in class.

Comments about the Problems

21. Homework This problem may be assigned as homework. Students may make top-view or side-view drawings or use words to explain why the captain has a better chance of seeing the swimmer by traveling in a zigzag course. Students should refer to vision lines and blind spots in their answers. If they have difficulty, you might have them model the situation with a toy boat. Have students use string to model the vision line.

Activity

For this activity, each group of students needs a piece of string and a toy boat. The boat can be made of either plastic or wood, but it must have a flat bottom.

Line up all the boats in the front of the classroom. For each boat, assign a number and determine the captain's location.

22. Without measuring, decide which boat has the largest blind spot and which has the smallest blind spot. Explain your decisions.

When comparing blind spots, you have to take into account the size of the boat. A large boat will probably have a large blind spot, but you must consider the size of the blind spot relative to the size of the boat.

23. In your group, use the following method to measure your boat's blind spot:

i. Place your boat on the grid on **Student Activity Sheet 7.** Trace the bottom of the boat. Attach a piece of string to the boat at the place where the captain is located. (The string represents the captain's vision line.)

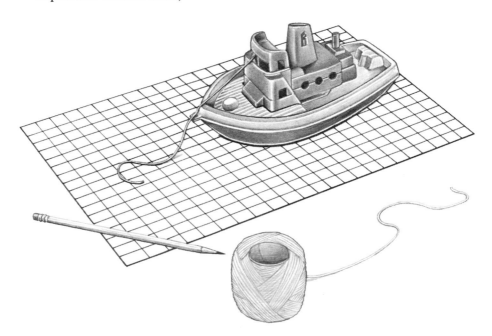

22. Estimates and explanations will vary. Students may base their explanations on the following features:

- The position of the bridge—a captain at the back of the boat will probably see less.

- The height of the bridge—a high bridge will improve visibility.

- The height of the bow or side of the boat—if the obstacle blocking the captain's view is high, the captain will see less.

- Some boats may have other obstacles besides the bow that block the captain's view.

23. Measurements will vary. Students can follow the steps on pages 12 and 13 of the Student Book unless they have decided to use another method. Students may also find the area of the bottom of each boat.

Materials Student Activity Sheet 7 (several copies per group); string (one roll per group); scissors (one pair per student); tape (one dispenser per group); toy boats with flat bottoms (one per group)

Overview Students estimate the relative sizes of the blind spots of different boats. They check their estimates by measuring the blind spots and draw conclusions based on their findings.

Planning Students may work on problems **22** and **23** in small groups. The activity will take an entire class period. The activity will work better if students have different-sized boats. If boats are not available, try using cars or other toy vehicles.

Comments about the Problems

23. Before beginning this problem, make sure that students understand what a blind spot is: the area of water that the captain cannot see. You may wish to discuss with students whether the water underneath the boat should be considered part of the blind spot as well. (In the steps listed on page 13 of the Student Book, the area underneath the boat is not considered part of the blind spot.)

Also, it must be clear in what direction the captain is looking. It is possible to have the captain look in all directions, but you may want to narrow this down to have the captain look straight ahead and sideways only (the steps listed on page 13 of the Student Book are based on the assumption that the captain looks only straight ahead and to the side).

ii. Using the string and a pencil, mark the spot on the grid where the captain's vision line hits the water. Make sure the vision line is taut and touches the edge of the boat.

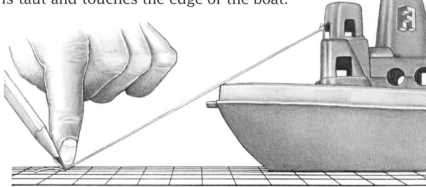

iii. Mark several places on the grid where the captain's vision line hits the water, so that you can determine the shape of the blind spot (the captain looks straight ahead and sideways). If the grid paper is not large enough, tape several pieces together. Draw the blind spot on the grid paper.

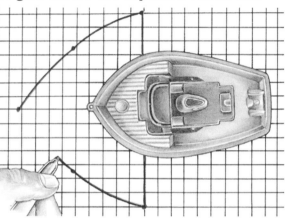

iv. Find the area of the blind spot. (*Note:* Each square of the grid is one square centimeter.)

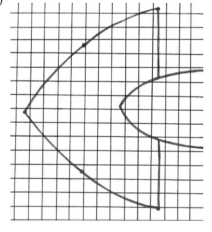

24. Make a list of the data for each boat. Decide which boat has the largest blind spot relative to its size and which has the smallest blind spot relative to its size.

Solutions and Samples
of student work

24. Tables and rankings will vary. Students should give their data in square centimeters. Sample table:

Boat	Area of the Blind Spot (in cm^2)	Area of the Boat (in cm^2)
1	55	44
2	91	68
3	94	64
4	56	192
5	844	176
6	140	136
7	237	60
8	754	56

In the above sample table, boats 1 and 4 have blind spots that are similar in size, students should look at the size of the boat. Boat 1 is much smaller, which means it has a relatively larger blind spot.

Hints and Comments

Materials Student Activity Sheet 7 (several copies per group); string (one roll per group); scissors (one pair per student); calculators (one per student); toy boats with flat bottoms (one per group)

Overview Students find the areas of blind spots for toy boats. They make a list of the data for the boats and draw conclusions about the sizes of the blind spots relative to the sizes of the boats.

Planning Students may work on problem **24** in small groups. Discuss students' results.

Comments about the Problems

24. Students should eventually realize that they must make a fair comparison. They must determine which boat has the largest or smallest blind spot relative to its size. Here are two possible ways:

Strategy 1

Divide both the area of each boat's blind spot and the area of the boat by the area of the boat. As a result, the area of the boat becomes 1, making it possible to compare the boats fairly, as shown below.

Boat	Area of Blind Spot (in boat units)	Area of Boat (in boat units)
1	1.25	1
2	1.34	1
3	1.47	1
and so on		

Encourage students to use a calculator. Discuss with students how many decimal points they need and how to round off.

Strategy 2

Look at the ratio between the area of the blind spot and the area of the boat as shown above. For instance, for boat 1 the ratio is 1.25:1. In other words, the area of the blind spot is 1.25 times as big as the area of the boat. This approach is basically the same as the first, but the reasoning is slightly different.

Activity

In this activity, you will investigate the blind area of a tugboat.

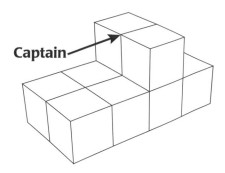

Captain

i. Build a model of the tugboat shown on the left with 1-centimeter blocks.

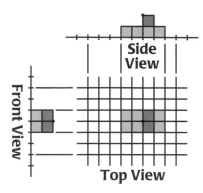

Side View

Front View

Top View

ii. Place your boat on the top-view outline of the tugboat on **Student Activity Sheet 8.**

iii. Use string to represent the captain's vision line.

25. a. On **Student Activity Sheet 8,** draw the captain's vision lines for the side, top, and front views.

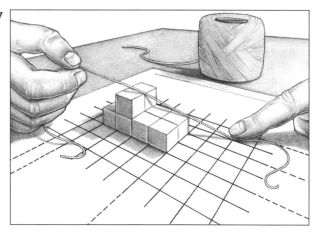

b. In the top view, shade the area of the grid that represents the blind area of the boat.

26. On **Student Activity Sheet 9,** draw vision lines and shade the blind area for the view shown. (One vision line has already been drawn.)

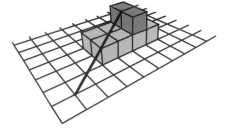

25.

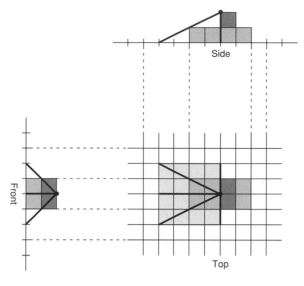

Side

Front

Top

26.

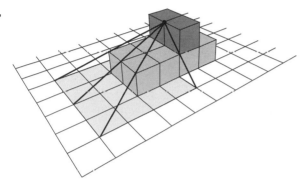

Materials Student Activity Sheets 8 and 9 (one of each per student); pieces of string, straws, or uncooked spaghetti (one per pair of students); centimeter cubes (at least 10 per pair of students)

Overview Students build a model of a tugboat and find the blind area for the tugboat's captain.

About the Mathematics Students delineated the blind area of boats in the previous activity, but this activity focuses their attention on the act of delineation itself and on the shape of the blind area.

This activity forms a bridge between the canyon tables activity—where the marks on the opposite wall were all on the same level—and the next section, where they will find that the shadows of objects of the same height standing in line will all end on the same line. The shape of a shadow is similar to the shape of the object casting the shadow, just like the shape of the blind area of a boat is similar to the boat itself.

Planning Students may work on problems **25** and **26** in pairs. The Extension below may be assigned as homework.

Comments about the Problems

25. Make sure students are able to draw vision lines correctly. If not, you might have students look carefully at their tugboats from the points of view depicted on Student Activity Sheet 8. While one student watches, his or her partner could model the captain's vision line with string, a straw, or a piece of uncooked spaghetti. Discuss the kind of information that is captured and lost in each view. In the side view, students see where the vision lines hit the water in front. In the front view, they see where the vision lines hit the water at the sides. In the top view, they see the directions of the vision lines.

26. Problem **26** builds on problem **25.** Students should be able to use the information they gathered in problem **25** to shade in the blind area.

Extension You may want to redesign the cube boat and have students delineate the blind area of the new boat.

Cars and Blind Spots

The above photograph is of a 1958 Pontiac Star Chief. This car is 5.25 meters long. Shown below is a side view of the car with vision lines indicating the blind area.

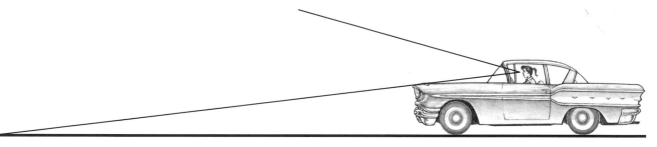

Today, cars are designed so that the blind area in front of the car is much smaller. The car shown below is a 1997 Buick Skylark that is 4.7 meters long. Notice how the vision line touches the hood of this car.

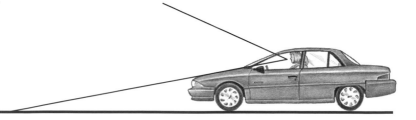

27. Find the length of road in front of each car that cannot be seen by the driver.

28. Which car has the longest relative blind spot?

29. What does the vision line that extends upward from each car indicate? Why is it important that this vision line be as close to vertical as possible?

27. Estimates will vary. Students may use a variety of strategies to find their answers. Sample strategy:

Cut a strip of paper and use it to mark the length of the Pontiac, bumper to bumper. The length of the Pontiac is 5.25 meters, so the length represented on the strip of paper must also be 5.25 meters. The strip fits in the length of the blind area almost two times. Therefore, the length of the blind area must be about 10 meters, as shown below.

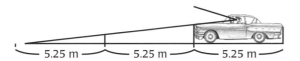

— 5.25 m — 5.25 m — 5.25 m —

Mark the length of the Buick on another strip of paper. The length represented on the strip is 4.7 meters. The strip fits in the length of the blind area less than once. So, the length of the blind area is about 3.5 meters, as shown below.

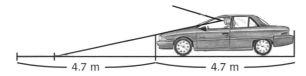

— 4.7 m — 4.7 m —

28. The Pontiac. The blind area of the Pontiac is about twice as long as the car. The blind area of the Buick is less than the length of the car. So, the Pontiac has a relatively longer blind spot.

29. Answers will vary. Sample response:

This vision line indicates what the driver can and cannot see when looking up. It is important that this line be as steep as possible so that the driver can see high traffic lights.

Materials scissors (one pair per student); $8\frac{1}{2}$" × 11" paper (one sheet per student)

Overview Students compare the lengths of the blind areas of two different cars: a classic car and a newer car. They also consider the importance of a large vision angle for the driver of a car.

About the Mathematics The idea of a vision angle is not made explicit here, but it is implied in problem **28.** You can refer to the idea that there are an infinite number of vision lines you can draw between those that define the field of vision. The crucial vision lines for the car indicate what the driver can see in front of the car, while looking up as well as down. Both vision lines form an angle.

Planning Students may work on problems **27–29** individually. These problems are optional. If time is a concern, you may omit these problems or assign them as homework. These problems can also be used as informal assessments. Discuss students' answers in class.

Comments about the Problems

27–29. Homework These problems may be assigned as homework.

27. Informal Assessment This problem assesses students' ability to measure blind spots and use ratios to solve problems involving steepness. Students should make a fair comparison by relating the length of the blind area to the length of the car.

28. Informal Assessment This problem assesses students' ability to make relative comparisons involving steepness problems.

29. Informal Assessment This problem assesses students' ability to understand the concept of a vision angle.

Bringing Math Home You might have students measure the length of the blind area of their family car, a bicycle, or a motorcycle. All the data can then be compared in class.

Summary

When an object is hidden from your view because something is in the way, the area that you cannot see is called the *blind area* or *blind spot.*

Vision lines are imaginary lines that go from a person's eyes to an object. Vision lines show what is in a person's line of sight, and they can be used to determine whether or not an object is visible.

In this section, you used vision lines to discover that the Colorado River is not visible in some parts of the Grand Canyon. You also used vision lines to find the captain's blind area for ships of various sizes.

Summary Questions

30. Describe a situation from your daily life which involves a blind spot. Include a picture of the situation with the blind spot clearly indicated.

30. Answers will vary. Sample student responses:

If I want to see the cat, I cannot see it if I'm standing behind the chair, because the chair is between us.

When I'm working at my desk, and I look sideways, there is a large area of the floor I cannot see.

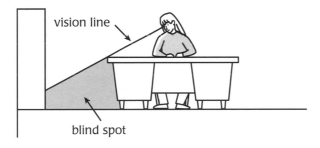

vision line

blind spot

When I wake up in the morning, I cannot see the floor next to my bed because the edge of my bed is blocking my view.

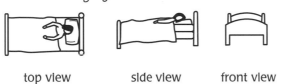

top view   side view   front view

Overview Students read the Summary, which reviews the main concepts covered in this section. They describe a situation from their daily lives involving a blind spot.

Planning Students may work on problem **30** individually. This problem may also be used as an informal assessment. You may want to discuss students' work in class. After students complete Section A, you may assign appropriate activities from the Try This! section, located on pages 47–50 of the *Looking at an Angle* Student Book, as homework.

Comments about the Problems

30. Informal Assessment This problem assesses students' ability to understand the concepts of vision line and blind spot, to construct vision lines and blind spots in two- and three-dimensional representations, and to choose appropriate views to draw situations involving steepness.

Check students' work for appropriate mathematical language, appropriateness of the pictures, and originality of the chosen context. In most cases, a top-view drawing will not be appropriate because it shows in which direction you are looking, but does not indicate the area you cannot see. The last example in the solutions column shows the drawings of a student who is struggling with this concept. The student added the side and front views only after feedback from the teacher.

Bringing Math Home This is a good opportunity for students to explain to their family how they can use mathematical language and drawings to describe real-world situations involving blind spots.

Writing Opportunity You may ask students to write their answers to problem **30** in their journals.

Work Students Do

In this section, students use the idea of blind spots to think about where to hide in a game of "Hide and See," a game similar to "Hide and Seek." They explore the differences between parallel and perspective projections of the game area. Students see the similarity between blind spots and shadows caused by a nearby light source. They also discover the differences in the shadows produced by a nearby light source and those produced by the sun. Students make top- and side-view drawings that show how shadows are projected by the sun. They measure the lengths of shadows produced by the sun's rays at different times of the day.

Goals

Students will:

- construct vision lines and blind spots (or light rays and shadows) in two- and three-dimensional representations;
- measure blind spots (or shadows);*
- measure angles;*
- understand the difference between shadows caused by a nearby light source and shadows caused by the sun;

* These goals are assessed in other sections of the unit.

- understand the ratio between an object and its shadow caused by the sun for different times of the day and the year;
- understand the relationship between steepness, angle, and height-to-distance ratio;
- choose appropriate views (top, side, or front) to draw situations involving steepness;*
- understand the correspondences between contexts involving steepness that may be represented with a right triangle.

Pacing

- approximately four 45-minute class sessions

About the Mathematics

When a vision line or the sun's rays hit an opaque object, a blind spot or a shadow is formed. A side view of this situation shows a right triangle having the vision line or light ray as the hypotenuse.

The vertical object causing the shadow is the one side and the shadow is the base or other side. Differences in the angle of the sun's rays or the height of the object produce differences in shadow length.

Shadows (blind spots) produced by a nearby light source (or the eye) fan out in different directions. When the light source is the sun, the shadows are parallel and the lengths of the shadows depend on the height of the sun and the angle between the sun's rays and the earth. This difference is not obvious to students, but with the experience they gain in this section, they can begin to understand the difference.

Side View

Materials

- Student Activity Sheets 10–16, pages 126–132 of the Teacher Guide (one of each per student)
- See the Hints and Comments column on each right-hand page of Section B for a complete list of the materials and quantities needed.

Planning Instruction

This section begins with a description of a game of "Hide and See," a game similar to "Hide and Seek." You may want to begin by asking students whether they have ever played Hide and Seek. Ask students, *How would you find a hiding place if it was not possible to hide inside or under anything?* [If it is not possible to hide inside a closet or under a bed, the next best hiding place may be in the seeker's blind spot, behind something.]

Students may work on problems 1–4 individually or in small groups. Problem 5 can be a whole-class activity. Students may work on problems 7–11 and 19–24 in small groups. The remaining problems may be done individually.

Problems 11, 16, 17, and 18 are optional. If time is a concern, you may omit these problems or assign them as homework.

Homework

Problems 1 (page 40 of the Teacher Guide), 2 and 3 (page 42 of the Teacher Guide), 4 (page 44 of the Teacher Guide), 12 (page 52 of the Teacher Guide), 13–15 (page 54 of the Teacher Guide), and 26 (page 62 of the Teacher Guide) may be assigned as homework. The Extension activities (pages 47 and 55 of the Teacher Guide), the Writing Opportunity (page 55 of the Teacher Guide), and the Bringing Math Home activity (page 53 of the Teacher Guide), may also be assigned as homework. After students complete Section B, you may assign appropriate activities from the Try This! section, located on pages 47–50 of the *Looking at an Angle* Student Book. The Try This! activities reinforce the key mathematical concepts introduced in this section.

Planning Assessment

- Problems 13–15 can be used to informally assess students' ability to understand the difference between shadows caused by a nearby light source and shadows caused by the sun and understand the ratio between an object and its shadow caused by the sun for different times of the day. They also assess students' ability to understand the relationship between steepness, angle, and height-to-distance ratio.

- Problems 17 and 22 can be used to informally assess students' ability to understand the ratio between an object and its shadow caused by the sun for different times of the day and the year. Problem 17 also assesses their ability to construct shadows in two- and three-dimensional representations.

- Problem 25 can be used to informally assess students' ability to understand the correspondences between contexts involving steepness that may be represented with a right triangle.

- Problem 26 can be used to informally assess students' ability to understand the difference between shadows caused by a nearby light source and shadows caused by the sun.

B. SHADOWS AND BLIND SPOTS

Hide and See

"Hide and See" is a game played by two teams, a red team and a blue team. The red team is positioned at the fort. The blue team is trying to approach the fort without being seen.

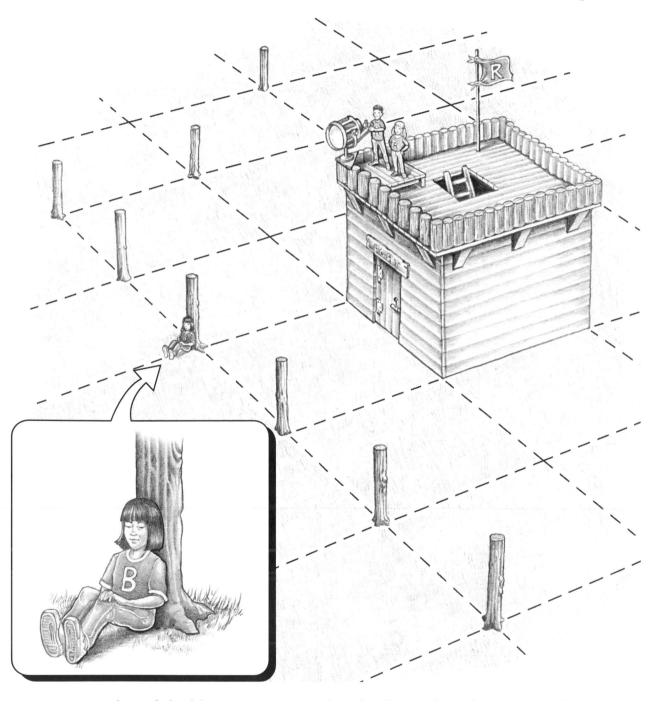

Regina, a member of the blue team, approaches the fort without being seen. She then sits behind a tree stump, unseen from the fort, as shown in the above picture.

1. Explain why the red team members on top of the fort cannot see Regina.

1. Explanations will vary. Sample student explanations:

The red team will not be able to see Regina, because Regina is in their blind spot.

The tree stump is blocking the red team's view.

The red team cannot see through the tree stump.

If you drew a vision line from the fort over the top of the tree stump, Regina would be sitting under the vision line, in the blind spot.

Students may draw a side view, as shown below.

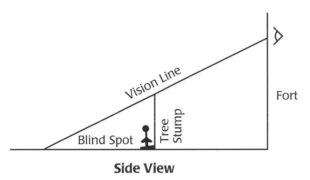

Side View

Overview Students are introduced to the game "Hide and See." They explain why a blue team member hiding in a certain position cannot be seen by members of the red team.

About the Mathematics A blind spot (or blind area) is a region not seen by someone; a shadow (or shaded area) is a region not reached by light rays. In both cases, there is an object that is blocking the "view." In the context of "Hide and See," the vision line cannot go through a tree stump, and neither can light rays.

Side-view drawings of situations involving blind spots and shadows are also similar. They look like a right triangle, as shown below.

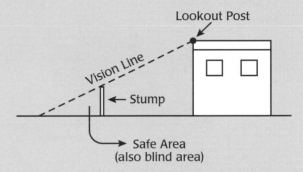

This underlying mathematics is not initially made explicit. The section is designed so that students will gradually learn to make these connections.

Students should be familiar with side- and top-view drawings from the unit *Side Seeing*.

Planning Students may work on problem **1** individually or in small groups. This problem may also be assigned as homework.

Comments about the Problems

1. Homework This problem may be assigned as homework. Based on their experiences in Section A, students should be able to identify the blind spot in this situation. Students may describe the location of the blind spot using words or drawings.

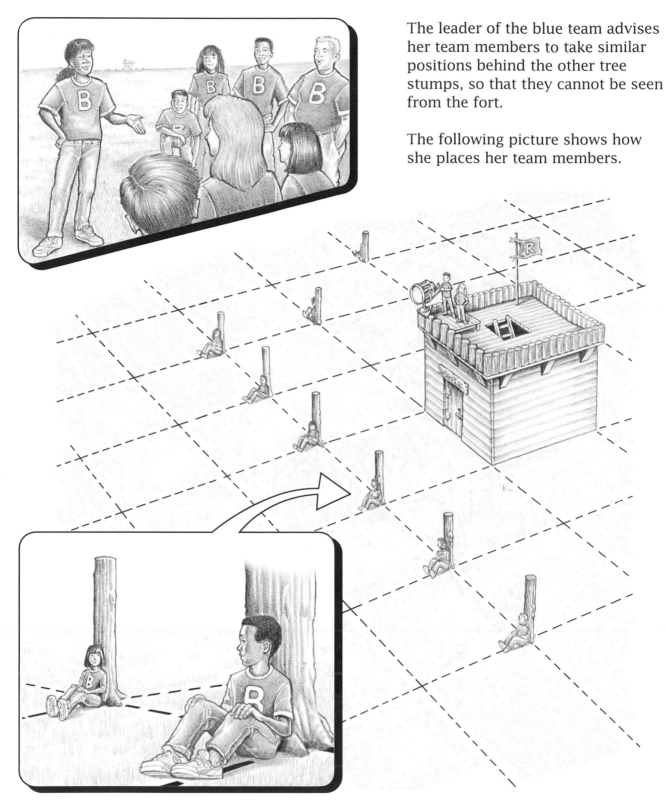

The leader of the blue team advises her team members to take similar positions behind the other tree stumps, so that they cannot be seen from the fort.

The following picture shows how she places her team members.

2. Can the members of the red team on top of the fort see any members of the blue team who are positioned behind the tree stumps? Explain.

3. Explain how each member of the blue team should sit so that they cannot be seen from the top of the fort. Include a sketch with your explanation.

2. Yes, the red team can see part of everyone except Regina. Explanations will vary. Sample student explanation:

   Only Regina's legs are pointing in the same direction as the vision line; everybody else's legs are pointing in a different direction.

   Students may try to draw a top or a side view. However, a side view may not really help because team members are sitting in a line. From the top, you can see that only one team member is hidden by the tree stump, as shown below.

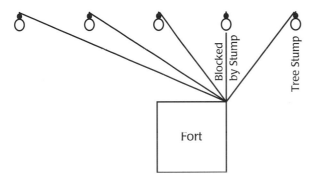

Top View

3. Explanations will vary. Sample student explanations:

   The legs of the blue team members should be pointing in the same direction as the vision lines. They should be fanning out.

   A top-view sketch is most appropriate in this case, since it shows the directions of the vision lines and the team members' legs.

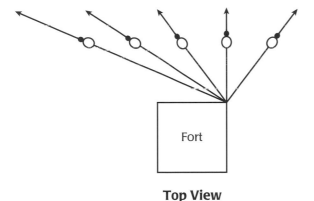

Top View

Materials straightedges or rulers (one per student); aluminum cans, optional (six cans of the same size per class); pencils, optional (six per class)

Overview Students continue solving problems within the context of a game of "Hide and See." They determine which blue team members may be seen by members of the red team.

Planning Students may work on problems **2** and **3** individually or in small groups. These problems may also be assigned as homework.

Comments about the Problems

2–3. Homework These problems may be assigned as homework.

2. Students may explain their answers using words or drawings. They might imagine vision lines from the fort to the stumps, or line up a pencil accordingly. This will lead them to the solution: the legs of the team members who are farther away are not in the blind spot. If students have difficulty, you might want to model the situation using a row of cans with pencils hidden behind them. Some students may have difficulty using words to explain why some team members can be seen. Encourage students to try to explain their thinking in words. You might want to have the class collaborate on a written solution.

The blue team members still are not quite sure how to position themselves behind the tree stumps without being seen. Regina has an idea. She suggests that they wait until night when the searchlight at the fort is turned on. She says, "We have to place all our team members in the shadows of the tree stumps made by the searchlight."

4. How will the searchlight help the blue team figure out how to sit behind the tree stumps without being seen?

The following pictures show the shadows made by the searchlight for two tree stumps. One picture is drawn using a perspective projection, the other is drawn using a parallel projection.

Nighttime Perspective Projection

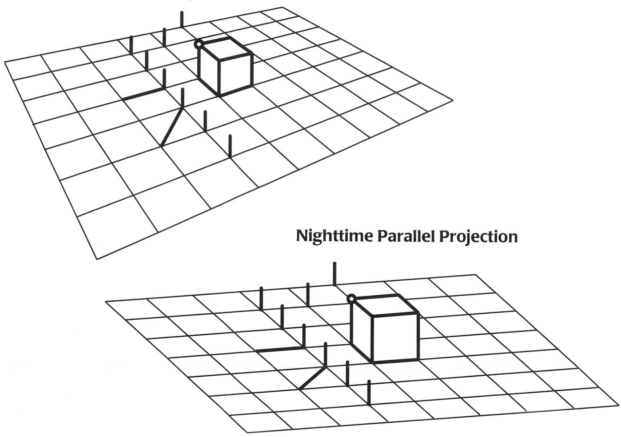

Nighttime Parallel Projection

5. Explain at least two differences between each projection.

6. On **Student Activity Sheet 10,** draw the shadows of the other tree stumps for each projection.

4. Answers will vary. Sample response:

If you turn on the searchlight, the shadows of the tree stumps will fan out, just as the vision lines and the light rays do. Blind spots are similar to shadows. Where your sight does not reach, the light does not reach either. The shadows will tell you where the blind spots are. The light is very close to the red team, in the corner of the fort. Ideally, it would be at exactly the same spot as the red team members' eyes.

5. Answers will vary. Sample response:

Perspective Projection: The squares of the grid that are closer are drawn bigger than the squares that are farther away. Objects that are the same length in reality have different lengths in the drawing. The drawing also has depth. It seems more realistic than the parallel projection.

Parallel Projection: Lines that are parallel in reality are parallel in the drawing. All objects that have the same lengths in reality also have the same lengths in the drawing. All squares of the grid are drawn equally large. The drawing does not look very realistic. This type of drawing is easier to work with if you want to measure or draw something.

6.

Nighttime Perspective Projection

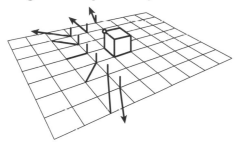

Nighttime Parallel Projection

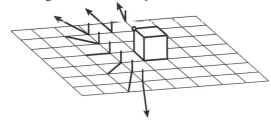

Materials Student Activity Sheet 10 (one per student); straightedges or rulers (one per student)

Overview Students explore the differences between perspective and parallel projections. They draw the shadows of tree stumps with a given light source.

About the Mathematics Both perspective and parallel projections have advantages and disadvantages for drawing three-dimensional reality. The *perspective projection* resembles what the eye perceives as real and the way a camera takes a photograph—the two tracks of a railroad seem to meet at the horizon and the parallel lines in the grid come closer to each other as they move farther away from the eye, toward the "point of convergence." However, in a technical or mathematical drawing, parallel lines never meet. This is accomplished by using *parallel projection*, although this looks unrealistic.

Planning Students may work on problems **4** and **6** individually. Problem **4** can also be assigned as homework. You may wish to have the whole class work on problem **5** together.

Comments about the Problems

4. Homework This problem may be assigned as homework. Students should begin to see the relationship between blind spots and shadows. What the eye does not see, a light would not reach either.

5. You may wish to briefly discuss some of the differences between the two projections and the advantages and disadvantages of each.

6. Any sketch showing the idea of the shadows fanning out is fine here.

Interdisciplinary Connection Perspective drawings show a scene as it would appear to the eye, not as it actually is. For example, in a perspective drawing, parallel lines appear to converge at a point far in the distance. The earliest artists to use a method of perspective drawing were probably the ancient Chinese. Artists in western Europe began to use perspective techniques during the Renaissance. You may want to have students try making their own perspective drawings of objects in the classroom or at home.

WITHDRAWN
CURRICULUM LIBRARY
GOSHEN COLLEGE LIBRARY
GOSHEN, INDIANA

For problem **6,** you made approximate drawings of the shadows of the tree stumps. However, to make a precise drawing of the shadows you need to know the following:

• the direction of the shadows, and

• the length of the shadows.

A side view and a top view of the shadows can help you to make a precise drawing.

Side View

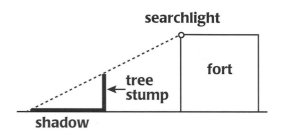

Top View

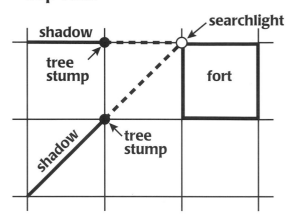

In the above drawings, the tree stump is half as high as the fort. The searchlight is at the top left corner of the fort.

7. How can the above pictures help you to make a precise drawing of the shadows?

8. On **Student Activity Sheet 11,** the shadows of two tree stumps have already been drawn in a picture of the top view of the fort and tree stumps.

a. Draw the shadows of the other tree stumps.

b. Explain how you found the directions and lengths for the shadows you drew.

Nighttime Top View

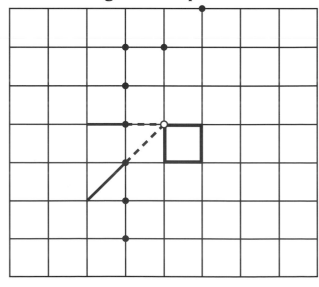

7. Answers will vary. Students should note that the top view tells you the directions of the shadows. The shadows extend the line that reaches from the light to a tree stump. However, a top view does not necessarily indicate where a shadow would stop. The side view allows you to determine where a light would hit the ground, which is where the shadow ends.

8. a.

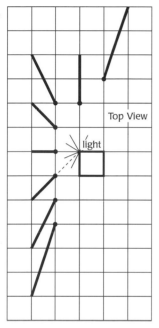

b. Explanations will vary. Sample explanation:

The length of each shadow is the same as the distance from the light to the tree stump. You can see this by comparing the two shadows that are given.

The shadows also form a pattern. All of the tree stumps that are on one line have shadows that end on one line. In addition, the shadows start one square from each other, and they end two squares apart. The directions of the shadows are also clear from the two shadows that are given. They fan out, just like the light. You can find the direction of each shadow by drawing a line from the light to the tree stump and extending it.

Materials Student Activity Sheet 11 (one per student); straightedges or rulers (one per student)

Overview Students learn how to make precise drawings of shadows using a top and a side view.

About the Mathematics If you want to make precise drawings of shadows, you need to know their length (which you can tell from the side view) and in what direction they go (which the top view shows). Note that in the side view, you can see only one tree stump; since the tree stumps are standing in line, all other tree stumps are hidden by the foremost one. The side view tells you that the length of each shadow is equal to the distance from the fort to that tree stump. You can also see this in the top view: the tree stumps are at different distances from the fort, but the shadow of a tree stump is still as long as the distance from the fort to the tree stump.

Knowledge about what kind of information is lost in a top or side view is important. In a top view, information about height or steepness is lost. In a side view, information about direction is lost. These concepts were informally introduced in the unit *Side Seeing*.

Planning Students may work on problems **7** and **8** in small groups. Discuss students' answers. You may want to have students return to problem **5** on page 19 of the Student Book to make adjustments to the perspective drawing.

Extension You may want to have students draw more tree stumps and their shadows on Student Activity Sheet 11.

Now back to the game "Hide and See." At night, the red team turns on the fort's searchlight. All the blue team members are sitting in the shadows of the tree stumps and cannot be seen from the fort.

At daybreak, the shadows of the tree stumps are caused by the sun, not the searchlight. The shadows have changed, so now the blue team members can be easily seen by the red team members on top of the fort.

Daytime Perspective Projection

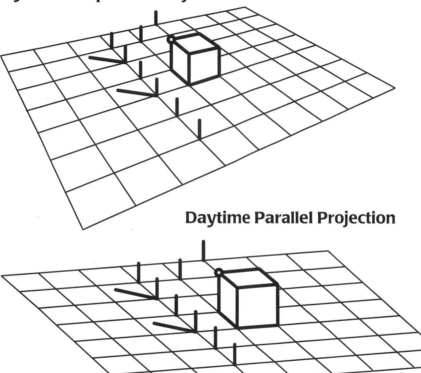

Daytime Parallel Projection

Daytime Top View

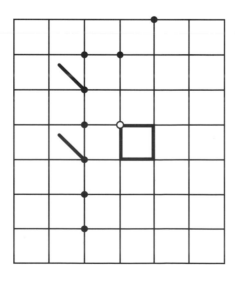

In the three pictures shown on the left, the shadows of two tree stumps caused by the sun have been drawn.

9. Draw the shadows of the other tree stumps on **Student Activity Sheet 12.**

10. Explain the differences between the shadows caused by the searchlight at night and the shadows caused by the sun during the day.

11. Draw the shadow of the fort in the top view on **Student Activity Sheet 12.** (*Note:* The fort is twice as tall as the tree stumps.)

Solutions and Samples
of student work

9. **Daytime Perspective Projection**

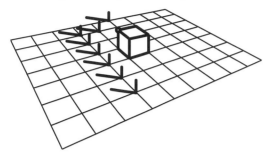

Daytime Parallel Projection

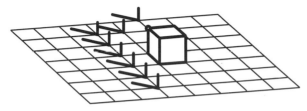

Daytime Top View

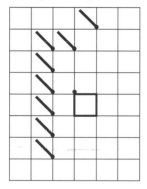

10. Explanations will vary. Sample explanation:

The shadows caused by the searchlight fan out. The shadows caused by the sun are of equal lengths, point in the same direction, and are parallel. This is because the sun is far away, while the searchlight is nearby.

11. **Daytime Top View**

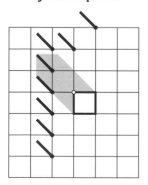

Materials Student Activity Sheet 12 (one per student); straightedges or rulers (one per student); $8\frac{1}{2}'' \times 11''$ paper, optional (one sheet per student); cardboard box, optional (one per class); string (one roll per class); scissors (one pair per class)

Overview Students discover that the shadows caused by the sun are parallel. They draw shadows caused by the sun.

About the Mathematics Shadows caused by the sun are different from shadows caused by a searchlight. Both the direction and the lengths of the shadows are different. Shadows caused by the sun do not fan out but are parallel, and the lengths of the shadows depend on the height of the sun and the time of year.

Planning Students may work on problems **9–11** in small groups. Discuss their answers. Problem **11** is optional. If time is a concern, you may omit this problem or assign it as homework.

Comments about the Problems

9. Encourage students to reason based on the two shadows that are given. If students have difficulty, you might have them draw the shadows in the top view first and then add them to the parallel and perspective projections.

11. Problem **11** is challenging. If students have difficulty, you might have them make a measuring strip out of a piece of paper and use it to measure the length of the shadow. Some students may be surprised that the shadow will cover tree stumps and their shadows. If it is sunny outside, have students look out the window. Some objects may be in the shadow of the school building.

Drawing the tip of the shadow of the fort correctly is the most difficult part. You may want to model the shadow using a box as the fort and string as a sun ray. Students might reason that if you draw the shadow of the corner of the fort's roof, the sides of the shadow's corner must be parallel to and equally as long as the sides of the fort itself. If you draw and connect the three endpoints of the shadow, this shows that the tip of the shadow has a 90-degree angle.

Shadows and the Sun

The sun causes parallel objects to cast parallel shadows. For example, the bars of the railing in the photograph on the right cast parallel shadows on the sidewalk.

Activity

In this activity, you will go outside on a sunny day to investigate the shadows caused by the sun.

First, you need to assemble your angle measure tool (AMT). Cut out the figure on **Student Activity Sheet 13** along the solid lines. Make the first fold as shown below and glue the matching shaded pieces together. Continue to fold your AMT in the order shown below.

Fold 1

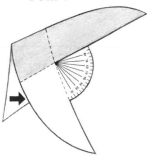

Fold 2

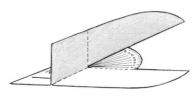

Fold 3

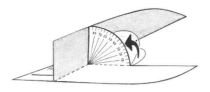

Fold 4

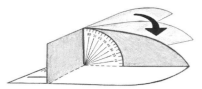

Students' angle measure tools should look like the one depicted in fold 4 on page 22 of the Student Book.

Materials Student Activity Sheet 13 (one per student); scissors (one pair per student); glue (one bottle per student); construction paper (one sheet per student)

Overview Students build an angle measure tool. They will use it to investigate the shadows caused by the sun. There are no problems on this page for students to solve.

About the Mathematics The underlying mathematical structure of the situation is again that of a right triangle: The light ray is the hypotenuse, the vertical object causing the shadow is the height, and the shadow is the base.

Planning Copy Student Activity Sheet 13 onto construction paper. Construction paper is sturdier than regular paper and will make a more accurate angle measure tool.

For your activity, you will need the following items:

- a stick about 1.2 meters long
- a stick about 0.7 meter long
- a centimeter tape measure
- several meters of string
- your AMT
- a directional compass

Put both sticks into the ground, about 2 meters apart. The longer stick should have a height of 1 meter above the ground, and the shorter stick should have a height of 0.5 meter above the ground. The sticks should be perfectly vertical.

12. In your notebook, copy the following table. Measure at least five different times during the day and fill in your table. (Add more blank rows to your table as needed.)

Time of Day	Direction of Sun	0.5-meter Stick		1-meter Stick	
		Length of Shadow (in cm)	Angle of Sun's Rays	Length of Shadow (in cm)	Angle of Sun's Rays

Use your compass to determine the direction from which the sun is shining. Use your tape measure to measure the lengths of the shadows of both sticks, and use your AMT and string (as shown below) to measure the angle between the sun's rays and the ground for both sticks. Be sure to stretch the string to where the shadow ends and place your AMT there.

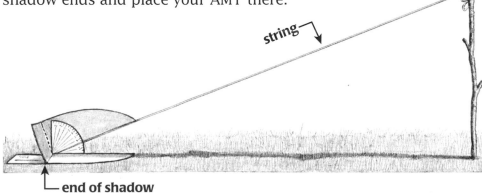

string

end of shadow

12. Answers will vary, depending on the number of measurements taken, the time of day, the season, the location, and so on.

Sample measurements:

Time of Day	Direction of Sun	0.5-meter Stick		1-meter Stick	
		Length of Shadow (in cm)	Angle of Sun's Rays	Length of Shadow (in cm)	Angle of Sun's Rays
11 A.M.	Southeast	50	45°	100	45°
4 P.M.	Southwest	100	27°	200	27°

Materials string (one roll per student); scissors (one pair per student); magnetic compasses (one per student); metric measuring tapes (one per student); sticks, 0.7 meter long (one per student); sticks, 1.2 meters long (one per student); angle measure tools constructed by students on page 22 of the Student Book (one per student)

Planning Students may do problem **12** individually. It may be assigned as homework over a weekend. Alternatively, you might plan with other teachers to have students take measures over a day's time.

Comments about the Problems

12. Homework This problem may be assigned as homework. The activity can be carried out only when the sun is shining: If it is impossible to do this experiment outside, either at school or at home, it can be replaced by a simulation using a strong light inside. However, you will not be able to model the sun exactly.

Students should use the sticks to create shadows. They should put the sticks in the ground in an open area, to avoid interfering shadows from surrounding objects. Students should use the string to model the sun's rays. To measure the angle of the sun's rays, students should place the angle measurement tool flat on the ground. The vertical flap should align perfectly with the string.

Some students may not know how to use a compass. The compass should be held flat, with the needle pointing in the direction of the sun.

The earlier in the day students start measuring, the lower the sun will be, the smaller the angle, and the longer the shadows. In other words, you will get a greater range of measures. Be sure to have students note the date and season that the measures were taken, as well as the location.

Bringing Math Home If students do this activity as homework, their families can participate. Students can discuss their findings at home.

Use your data from the table in problem **12** to answer the following problems:

13. a. Describe the movement of the sun during the day.

   b. Describe how the directions of the shadows change during the day. How are the shadows related to the direction from which the sun is shining?

   c. Describe the changes in the lengths of the shadows during the day. When are the shadows the longest, and when are they the shortest?

14. a. Compare the shadows of the longer stick with the shadows of the shorter stick. Describe the relationship between the length of the shadow and the height of the stick.

   b. Were the shadows of the two sticks parallel at all times? Explain.

15. a. Compare the angle of the sun's rays for each stick at any moment during the day.

   b. Describe how the angle of the sun's rays changed during the day. When is the angle the largest, and when is it the smallest?

   c. How is the size of the angle of the sun's rays related to the length of the shadows?

13. a. Descriptions will vary. Sample response:

The sun rises in the east and sets in the west. Also, in the Northern Hemisphere, the sun travels across the southern part of the sky.

b. Descriptions will vary. Sample response:

The direction of the shadows is exactly opposite the direction from which the sun is shining (a difference of 180°).

c. Descriptions will vary. Sample response:

The shadows are longest at the beginning and end of the day. They are shortest at noon.

14. a. Descriptions will vary. Sample response:

The shadow of the shorter stick is always half as long as that of the longer stick if you measure the shadows at the same time. So, at any given moment of the day, the ratio between an object's length and that of its shadow is constant.

b. Yes. Explanations will vary. Sample explanations:

If the sticks are vertical, the shadows will always be parallel.

The sun hits the sticks at the same angle.

15. a. The angle of the sun's ray is the same at any given moment, no matter how tall the object is.

b. When the sun is near the horizon, the angles are small. At noon, the angles are larger.

c. When the sun is low in the sky and the angle with a certain location on Earth's surface is small, then the shadows are long. When the sun is high in the sky and the angle with Earth's surface is larger, the shadows are shorter.

Overview Students analyze the data they collected in problem **12** and draw conclusions.

About the Mathematics In the unit *Ratios and Rates*, the ratio between the length of a stick and its shadow is investigated.

Planning Students may work on problems **13–15** individually. These problems may also be assigned as homework, and can be used as informal assessments.

Comments about the Problems

13–15. Informal Assessment These problems assess students' ability to understand the difference between shadows caused by a nearby light source and shadows caused by the sun and to understand the ratio between an object and its shadow caused by the sun at different times of the day. They also assess students' ability to understand the relationship between steepness, angle, and height-to-distance ratio. These problems can also be assigned as homework.

13. c. Students may give their answers in words, or they may make a graph with times of day on the *x*-axis and the lengths of the shadows on the *y*-axis.

Writing Opportunity You may want to have students write a report explaining their data and the results of the activity. Students should include data, drawings, tables, and graphs in their reports.

Extension You may want to challenge students by asking them the following questions:
- *How would the size of the angles of the sun be different during another season?* [Summer days are longer than winter days. Also, the sun is higher in the summer than in the winter. The angle of the sun's rays and the length of the shadow change accordingly.]
- *How would the size of the angle of the sun be different if you lived closer to the Equator?* [The sun is higher when you are closer to the Equator, so shadows are shorter and the angle of the sun's rays is larger.]
- *How would your measurements be different if you lived in the Southern Hemisphere?* [In the Southern Hemisphere, the sun travels across the northern part of the sky.]

The picture below shows the shadows of two buildings at noon. The sun is shining from the south. One building is twice as tall as the other.

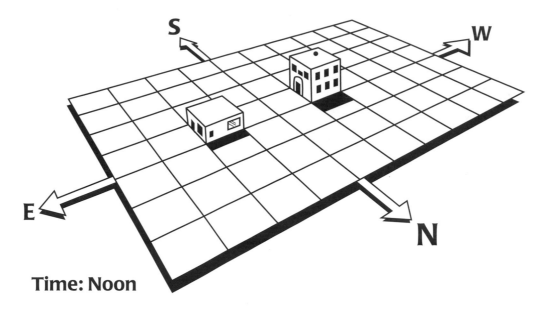

Time: Noon

16. Study the shadows of the buildings shown above. Describe the directions and the lengths of the shadows.

In the four pictures below, you see the two buildings at different times of day.

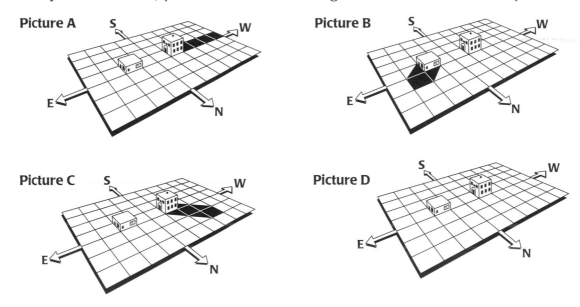

Picture A

Picture B

Picture C

Picture D

17. a. On **Student Activity Sheet 14,** draw the shadows that are missing. (*Note:* Picture D needs both shadows drawn in.)

   b. Label each picture with an appropriate time of day.

18. Describe how the lengths of the shadows change during the day. Make drawings or graphs to illustrate your descriptions.

16. Descriptions will vary. Sample description:

The sun is shining from the south, so the shadows fall toward the north. The shadow of the shorter building is half as long as the shadow of the taller building.

17. a.

Picture A	Picture B

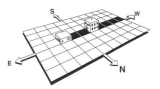 

Time: 7:00 A.M. (sunrise)   Time: 2:30 P.M. (mid-afternoon)

Picture C	Picture D

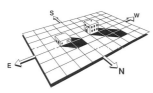 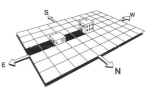

Time: 9:30 A.M. (mid-morning)   Time: 5:00 P.M. (sunset)

Students will give a variety of answers for the last drawing. If they choose sunset, the shadows fall in a direction opposite to the direction they fall at sunrise (Picture A).

b. The times of the day are shown above. The pictures are labeled for winter. The sun rises at 7:00 A.M. and sets at 5:00 P.M.

18. Descriptions will vary. Sample description:

Shadows are long at sunrise. They get shorter as the day progresses toward noon. At noon, the shadows are their shortest. After noon, they start getting longer, until sunset.

Drawings and graphs will vary. Sample graph:

Shadow Lengths throughout the Day

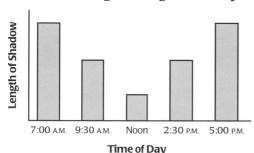

Materials Student Activity Sheet 14 (one per student); straightedges or rulers (one per student)

Overview Students reflect on how the direction and the length of shadows change during a day. They describe these changes.

Planning Students may work on problems **16–18** individually. If time is a concern, you may omit these problems. Problem **17** may be used as an informal assessment.

Comments about the Problems

16. Students should remember from the previous activity that at a certain time of the day there is a constant ratio: a building twice as tall as another building gives a shadow twice as long.

17. Informal Assessment This problem assesses students' ability to understand the ratio between an object and its shadow caused by the sun for different times of the day and to construct shadows in two- and three-dimensional representations.

b. Remind students that the sun rises in the east, sets in the west, and travels across the southern part of the sky.

18. Students may remember how to make graphs from the units *Picturing Numbers* and *Tracking Graphs*. If they do not yet know how to make graphs, let them try anyway. Any graph showing the change of the shadow lengths over time is acceptable.

The lower the sun is, the longer the shadows that are cast. The height of the sun not only depends on the time of the day, but also on the season. Shown below is a side view of a building around noon during the summer.

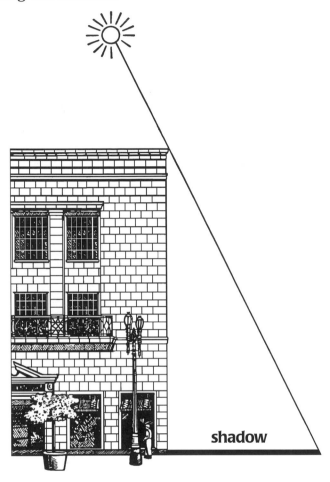

The length of the above building's shadow is one-half the height of the building.

19. Measure the angle between the sun's rays and the ground.

Around noon during the winter, the length of this building's shadow is $2\frac{1}{2}$ times the height of the building.

20. a. Draw a side view of the building and its shadow around noon during the winter.

   b. Measure the angle between the sun's rays and the ground.

Around noon during the spring, the angle between the sun's rays and the ground is 45°.

21. a. Draw a side view of the building and its shadow around noon during the spring.

   b. If the building is 40 meters tall, how long is its shadow?

22. Describe the changes in the length of the shadow and the angle of the sun's rays from season to season.

19. The angle is about 65°.

20. a.

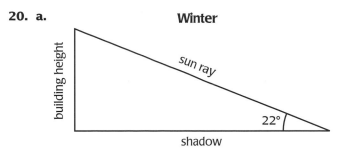

Winter

building height

sun ray

22°

shadow

b. The angle is 22°.

21. a.

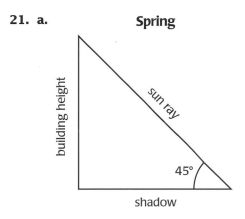

Spring

building height

sun ray

45°

shadow

b. The length of the shadow is equal to the height of the building, which is 40 meters.

22. Descriptions will vary. Students may make a table using all the information they have gathered, as shown below.

Season	Angle of Sun's Rays	Height of Building	Length of Shadow
Summer	65°	40 m	20 m
Winter	22°	40 m	100 m
Spring	45°	40 m	40 m

At noon on a summer's day, the sun can appear to be almost directly overhead, and shadows are very short. In the fall, the shadows gradually get longer and the angle of the sun's rays gradually gets smaller. In the winter, the noontime sun is relatively low and creates long shadows. In the spring, the angle of the sun's rays increases gradually, and shadows get shorter.

Materials straightedges or rulers (one per student); protractors and/or compass cards (one per student); graph paper, optional (several sheets per student)

Overview Students reflect on how the lengths of shadows and the angles of the sun's rays vary depending on the season. Students make side-view drawings of a building, its shadow, and the sun's rays for three seasons. The drawings are right triangles.

Planning Students may work on problems **19–22** in small groups. You may want to discuss these problems with the whole class. Problem **22** may be used as an informal assessment.

Comments about the Problems

19. Students should measure the angle with a protractor or compass card.

20. Students' drawings may be to different scales. In any case, the length of the shadow should be two and a half times the height of the building. You may ask students why drawings using different scales give the same angle measurement. (*Note:* The larger the drawing, the more precise the angle measurement will be.)

21. Again, students' drawings may be to different scales. Students should discover that with a 45° angle, the ratio between the height of the building and the length of the shadow is 1:1.

22. Informal Assessment This problem assesses students' understanding of the ratio between an object and its shadow caused by the sun for different times of the year.

Shadows and a Light

Now back to the game "Hide and See." Recall that the shadows cast by the fort's searchlight during the night were different than those cast by the sun during the day.

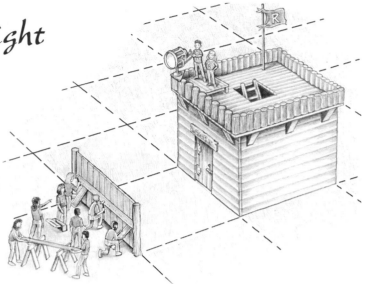

Suppose that the blue team builds a wall between the two tree stumps directly in front of the fort. The wall is the same height as the tree stumps.

The following picture shows the shadow of the wall when the searchlight is on.

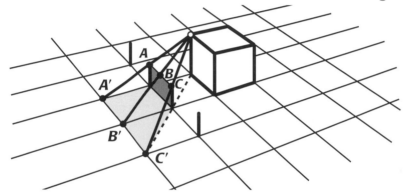

Point *A'* is the shadow of point *A*, point *B'* is the shadow of point *B*, and point *C'* is the shadow of point *C*.

The red team decides to move the searchlight to the other front corner of the fort.

23. a. On **Student Activity Sheet 15,** draw the shadow of the wall caused by placing the searchlight on the other front corner of the fort.

   b. Does moving the searchlight from one front corner of the fort to the other change the area of the shadow? Explain.

Suppose the wall is placed twice as far from the fort as shown on the right. The leader of the blue team thinks that this will make the area of the shadow caused by the searchlight twice as big, thus allowing more room for the blue team to hide.

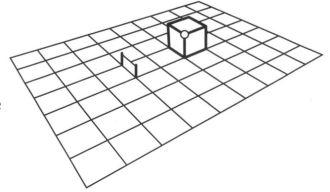

24. Is the leader of the blue team correct? Explain. (You might want to make a top-view drawing to find your answer.)

23. a.

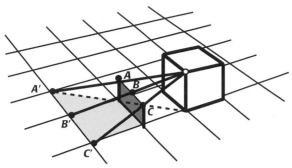

b. Moving the light does not change the area of the shadow. This is clear in top-view drawings, as shown below.

Top View

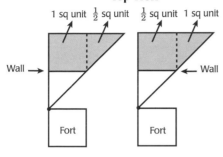

24. Yes, the area of the shadow will be twice as large. By doubling the distance between the light source and the wall, you also double the length of the shadow, thus doubling its area.

Top View

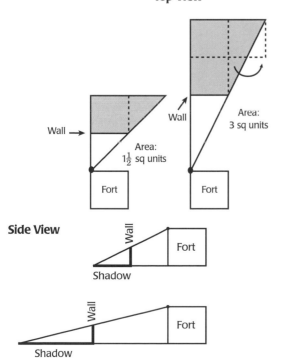

Materials Student Activity Sheet 15 (one per student); straightedges or rulers (one per student); graph paper, optional (several sheets per student)

Overview Students return to the game of "Hide and See." They construct shadows caused by a searchlight for a wall that is moved to different locations.

Planning Students may work on problems **23** and **24** in small groups.

Comments about the Problems

23. This problem is challenging because the drawing is in perspective. Students may use the example shown on page 27 of the Student Book as a guide when they construct the shadow with the searchlight on the other front corner.

You might have students draw the shadow when the searchlight is in the center of the wall of the fort. Ask them, *Does the area of the shadow change in this case?* [No.]

24. Some students may focus on the area of the shadow, in which case the top view is helpful. Others may look at the length of the shadow, which becomes visible in a side view. Ask students, *What happens to the size of the shadow as the wall is moved farther and farther away from the fort?* [The shadow gets longer.]

Summary

Shadows are similar to blind spots (or blind areas). Shadows can be caused by two kinds of light:

- light that is nearby, such as a streetlight

- light that is very far away, such as the sun

When light comes from a nearby point, shadows are cast in different directions. When light comes from a far away source, the light rays are parallel, and the shadows are also parallel.

Summary Questions

25. Describe what a shadow is. How are shadows and blind spots similar?

In the picture on the right, you see a streetlight surrounded by posts.

26. On **Student Activity Sheet 16,** draw in the missing shadows. In top view A, it is nighttime, and the streetlight is on. In top view B, it is daytime, the streetlight is off, and the sun is shining.

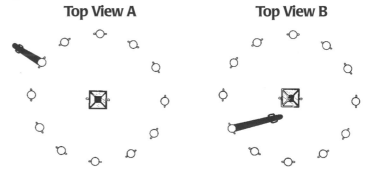

Top View A   Top View B

25. Descriptions will vary. Sample description:

A shadow is a space or an area that light rays do not reach. Shadows and blind spots are similar. A blind spot is a space or an area that your eye cannot see. Shadows and blind spots can both be represented with a drawing of a right triangle, as shown below.

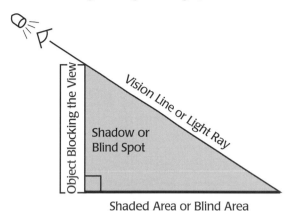

26.

Top View A

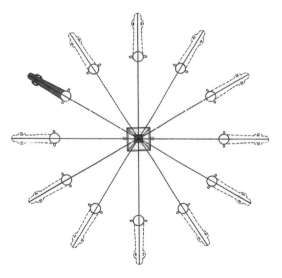

Top View B

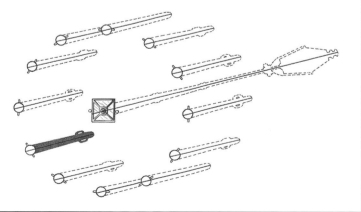

Materials Student Activity Sheet 16 (one per student); straightedges or rulers (one per student)

Overview Students read the Summary, which reviews the main concepts covered in this section. They reflect on what they have learned about the differences between shadows caused by a light and shadows caused by the sun.

Planning Students may work on problems **25** and **26** individually. These problems may also be used as informal assessments. Problem **26** can also be assigned as homework. Discuss students' answers in class. After students complete Section B, you may assign appropriate activities from the Try This! section, located on pages 47–50 of the *Looking at an Angle* Student Book, for homework.

Comments about the Problems

25. Informal Assessment This problem assesses students' ability to understand the correspondences between contexts involving steepness that may be represented by a right triangle.

Some students may make a drawing to illustrate their answers while others may use only words. Both approaches are fine, as long as students show that they understand the similarities.

26. Informal Assessment This problem assesses students' ability to understand the difference between shadows caused by a nearby light source and shadows caused by the sun. This problem can also be assigned as homework.

In the first picture, all shadows should be equally long, and they should be fanning out from the lamp. In the second picture, all shadows of the posts are also equally long, but they should be parallel to each other. To make their drawings complete, students should also include the shadow of the streetlight. To estimate the length of the streetlight's shadow, students should remember that at a given time of the day, the ratio between an object's length and the length of its shadow is constant.

Work Students Do

In this section, students describe how the shadows at Acoma Pueblo vary at different times of the day. They draw the shadows of the rungs of a ladder that is leaning against a wall and indicate the direction of the rays of sunlight that create the shadows. Then students investigate how changes in the ladder's steepness affect the triangle that is formed by the ladder, the ground, and the wall. They make side-view drawings of a ladder leaning against a wall in a variety of circumstances. Students then investigate and describe a ladder's steepness using the height-to-distance ratio and the angle between the ground and the foot of the ladder. They make a graph of the relationship between the height-to-distance ratio and the angle of elevation. Finally, students determine which of two vision lines in a canyon is steeper, and support their answers using the height-to-distance ratio and the angle of elevation.

Goals

Students will:

- understand the concept of steepness;
- measure angles;
- make scale drawings of situations involving steepness;
- make relative comparisons involving steepness problems;
- understand the relationship among steepness, angle, and height-to-distance ratio;
- choose appropriate views (top, side, or front) to draw situations involving steepness;*
- understand the correspondences between contexts involving steepness that may be represented with a right triangle;*
- use ratios to solve problems involving steepness.

  * These goals are assessed in other sections of the unit.

Pacing

- approximately three 45-minute class sessions

Vocabulary

- alpha
- beta
- gamma
- steepness

About the Mathematics

The steepness of a line can be described by using the ratio of change in height over change in distance (which is actually the same thing as the slope) or by using the angle of elevation (or inclination). For example, a line with a height-to-distance ratio of 1:1 has an angle of elevation of 45 degrees, as shown on the right.

The larger the height-to-distance ratio, the steeper the line.

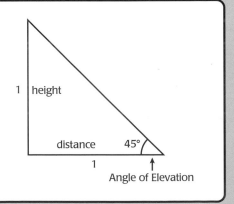

Materials

- Student Activity Sheet 17, page 133 of the Teacher Guide (one per student)
- rulers, pages 69, 71, 73, 75, and 77 of the Teacher Guide (one per student)
- books or boxes that can stand upright, page 71 of the Teacher Guide (one per student)
- ladder, pages 69, 71, and 75 of the Teacher Guide, optional (one per class)
- protractors or compass cards, pages 73 and 77 of the Teacher Guide (one per student)
- graph paper, pages 73 and 75 of the Teacher Guide, optional (two sheets per student)

Planning Instruction

You might introduce this section with a brief discussion about Acoma Pueblo, New Mexico. The village was built by Pueblo Native Americans. In 1540, Spanish conquistador Francisco Vásquez de Coronado described Acoma's location as the strongest defensive position in the world. Ask students, *If you wanted to build a city that would be easy to defend against attack, where would you build it and what would it be like?* [Answers will vary, but Acoma is defensible because it is located at the top of a butte, with cliffs all around it. In addition, the houses at Acoma originally did not have doors. People entered them by climbing ladders to the upper stories. In case of attack, the ladders could be removed.]

Students may work on problems 1–3 individually or in small groups. Problems 4 and 5 should be worked on as a whole class. Problems 6–9 and 14 may be done in small groups. Students may work on problems 10–13 and 15 individually.

There are no optional problems in this section.

Homework

Problems 11 and 12 (page 74 of the Teacher Guide) may be assigned as homework. After students complete Section C, you may assign appropriate activities from the Try This! section, located on pages 47–50 of the *Looking at an Angle* Student Book. The Try This! activities reinforce the key mathematical concepts introduced in this section.

Planning Assessment

- Problem 10 can be used to informally assess students' ability to measure angles and make scale drawings of situations involving steepness.
- Problem 15 can be used to informally assess students' ability to understand the concept of steepness, measure angles of steepness, and make relative comparisons involving steepness problems. It also assesses their ability to understand the relationship among steepness, angle, and height-to-distance ratio and use ratios to solve problems involving steepness.

C. SHADOWS AND ANGLES

Acoma Pueblo

The Acoma Pueblo is considered the oldest continually inhabited village in the United States. The above drawing is of the Acoma Pueblo as it might have looked over 100 years ago. Located near Albuquerque, New Mexico, it is famous for its beautiful pottery and architecture. By analyzing the pottery, archaeologists have determined that this village was settled about 1,000 years ago.

The photograph on the left shows the typical architecture of a main street of the village. This picture was taken in the morning.

1. Describe how the shadows will be different at noon.

1. Answers will vary, but students should note that the shadows will become shorter as noon approaches. At noon, the sun will be shining from a different direction. In the Northern Hemisphere, the sun shines from the east in the morning and from the south around noon.

Overview Students read about Acoma Pueblo. They study the shadows in a photograph of Acoma Pueblo.

Planning Students may work on problem **1** individually or in small groups.

Comments about the Problems

1. You might remind students of the activity on pages 22–24 of the Student Book, in which they measured shadows of sticks at different times of the day. In that activity, they learned that the sun shines from different directions at different times of the day and that the lengths of shadows change during the day.

Did You Know? Acoma Pueblo is also known as "Sky City." The pueblo's name comes from the Keresan language: *ako* means "white rock," and *ma* means "people." Acoma's people live in adobe houses at the top of a sandstone butte 357 feet (109 meters) high. Most make a living by farming on the plains below or by making pottery. The village has been inhabited since the 10th century. Logs for its mission church, San Esteban Rey, and dirt for its graveyard were hauled up the cliffs in 1629–1641. Today, people can reach Acoma by road or by climbing a staircase cut into the rock.

Originally, the houses in the Acoma Pueblo had no front doors; ladders were used to enter the houses on the second floor. Ladders propped against the houses formed different angles. The steepness of the ladders can be measured several ways.

Recall from Section B that the sun's rays are parallel. The drawing below shows a ladder and its shadow. The drawing also shows how the sun casts a shadow for one rung of the ladder.

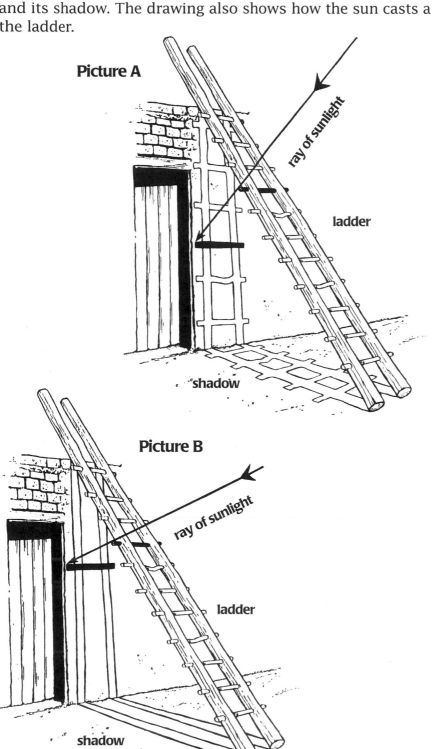

Picture A

ray of sunlight

ladder

shadow

Picture B

ray of sunlight

ladder

shadow

2. On picture A on **Student Activity Sheet 17,** draw a ray of sunlight that casts a shadow for each of the other 10 rungs.

The drawing on the left shows the same ladder in the same position, but at a different time of day.

3. On picture B on **Student Activity Sheet 17,** draw a ray of sunlight and the corresponding shadow for each of the other 10 rungs.

2.

Picture A

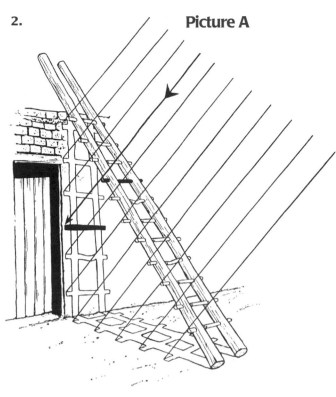

Materials Student Activity Sheet 17 (one per student); rulers (one per student); ladder, optional (one per class)

Overview Students study the shadows of the rungs of a ladder. They draw rays of sunlight for a situation with a ladder, with given shadows, and with one given ray of sunlight. Then they draw shadows and rays of sunlight for the same setting at a different time of day.

About the Mathematics The context of the ladders is used to bridge the subjects of shadows (from Section B) and steepness, which will be the main mathematical focus of this section.

Planning Students may work on problems **2** and **3** individually or in small groups.

Comments about the Problems

2. It should be clear to students that the sun's rays touch the tip of the rung and continue from there to the tip of the shadow of the rung. Remind students that shadows are created because light cannot pass through opaque objects (just as people cannot see through opaque objects). If students have difficulty, you might have them observe shadows made by a light shining on a real ladder.

3.

Picture B

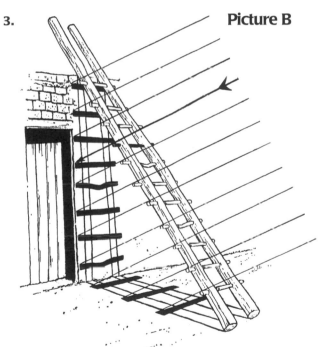

Did You Know? Observing the length and direction of a shadow is an ancient method of measuring time. As early as 2000 B.C., people used gnomons to track the movement of the sun across the sky throughout the day and the year. A gnomon (pronounced NOH mun) is a vertical object, usually a shaft inserted in the ground. People tracked the movement of the sun by observing changes in the gnomon's shadow. In fact, a sundial has a gnomon and a specially constructed hour scale. The gnomon of a sundial makes a shadow that falls on the hour scale and marks the time of day.

The following drawings show two side views of the same ladder leaning against a wall.

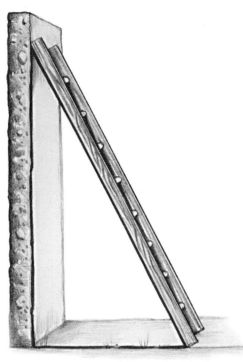

4. Describe the differences between the ways the ladder is positioned against the wall in the above drawings.

5. **a.** What problems might occur if the ladder is very steep?

   b. What problems might occur if the ladder is not steep enough?

As the steepness of the ladder changes, the following measures also change:

- the height on the wall that can be reached by the top of the ladder,

- the distance between the foot of the ladder and the wall,

- the angle between the ladder and the ground.

6. Investigate different levels of steepness by using a ruler or pencil to represent a ladder and an upright book or box to represent a wall. Describe your discoveries. (You may use drawings.)

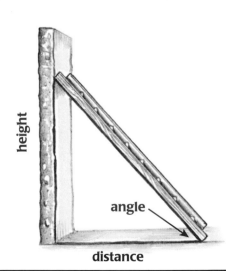

4. Answers will vary. Sample student response:

When I compared the two ladders, I noticed that the ladder on the left is positioned low on the wall and is not very steep. The ground distance that this ladder covers is much greater than that of the ladder on the right. The angle formed by the foot of this ladder and the ground is small. The ladder on the right is positioned high on the wall and is much steeper. The ground distance that this ladder covers is much less. The angle formed by the foot of the ladder and the ground is much larger.

5. a. Answers will vary. Sample response:

When the ladder is really steep, it may be dangerous. It may tip backwards when someone climbs it.

b. Answers will vary. Sample response:

A person climbing such a ladder may not be able to reach the top of the wall due to its low position on the wall.

When the ladder is not very steep, or too flat, it may collapse.

6. Answers will vary. Students may use pictures and tables to describe their findings. The sample response below shows a pattern among the pictures of the triangles, the ratios of height-to-ground distance, and the angles.

Pictures	Ratio (height:distance)	Angle
1 ◺ 5	1:5	11° or 12°
2 ◺ 4	2:4	26° or 27°
3 ◺ 3	3:3	45°
4 ◺ 2	4:2	63° or 64°
5 ◺ 1	5:1	78° or 79°

When the ladder is positioned higher on the wall, the distance that the ladder covers on the ground becomes shorter and the angle of the ladder to the ground becomes larger. When the ratios are inverses (for example, 1:5 to 5:1), the two angles add up to 90°.

Materials rulers, optional (one per student); books or boxes that can stand upright (one per student); ladder, optional (one per class)

Overview Students explore the steepness of a ladder leaning against a wall.

About the Mathematics The concept of steepness is informally introduced. Students learn that to measure steepness three things are involved: the position of the top of the ladder on the wall, the distance between the wall and the foot of the ladder, and the angle formed between the ladder and the ground. The concept of steepness will be formalized later in the section.

Planning You may want to discuss problems **4** and **5**. Students may work on problem **6** in small groups.

Comments about the Problems

4. At this point, students' answers need not be well developed. You may want to have a whole-class discussion in which students make a list of the differences they have found.

5. The answer will be obvious if you can demonstrate with a real ladder.

6. Encourage students to look more closely at the relationship between the height and ground distance, and between the ratio of the height-to-ground distance and the angle. This problem can be used to lead students to see the relationship between inverse ratios and their angles. For instance, the ratios 1:5 and 5:1 (and 2:4 and 4:2) have angles that add up to 90 degrees.

On the right is a drawing
of a ladder leaning against
a wall. Angles are often
given names. Sometimes
the name of the angle is
a letter of the Greek
alphabet. The first letter
in the Greek alphabet is
α (*alpha*), the second
letter is β (*beta*), and the
third letter is γ (*gamma*).

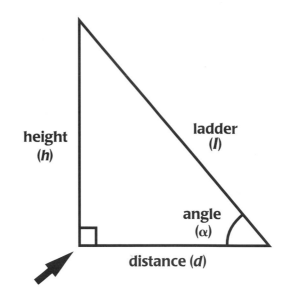

This symbol is used to
indicate a 90°, or right, angle.

7. Why must the angle between the height (*h*) and the distance (*d*) be 90°?

8. Measure the angle α for the ladder in the above drawing.

There are several ways to measure the steepness of a ladder. You can measure the angle α,
or you can find the ratio of the height to the distance. The ratio of the height to the
distance can be expressed as a fraction or a decimal.

9. What happens to the angle α as the ratio of the height to the distance increases?

10. Use a compass card or a protractor and a ruler to make side-view drawings to scale
of a ladder leaning against a wall for each of the following situations. Also, label
α, *h*, and *d* with their measures, and find the height-to-distance ratio.

    a. α = 45°

    b. *h* = 2, *d* = 1

    c. α = 30°

    d. *h* = 1, *d* = 2

    e. α = 60°

7. By definition, height is always measured vertically.

8. Angle α is equal to 50°.

9. As the ratio gets larger, the height increases, the ground distance decreases, and the angle α becomes larger.

10. a.

The ratio *h*:*d* is 1:1.

b.

The ratio is 2:1. Angle α is about 63°.

c.

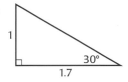

Using a scale drawing, the ratio *h*:*d* is about 4:7 or 1:1.7.

d.

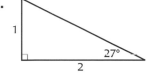

The ratio is 1:2 and is the inverse of the ratio in problem **b.** Since the angles of inverse ratios must add up to 90°, angle α is 90° − 63°, or 27°.

e.

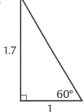

Angle α is 60°. Angle α for problem **c** is 30°. Together, they add up to 90°. That means their ratios are inverses. Since the ratio for problem **c** is 4:7, the ratio for this problem must be 7:4.

Materials rulers (one per student); protractors or compass cards (one per student); graph paper, optional (two sheets per student)

Overview Students measure heights, distances, and angles in side-view drawings of ladders and describe the relationships between these measurements. They also construct side-view drawings of ladders.

About the Mathematics The height-to-distance ratio may be expressed in different ways: as a ratio, as a fraction, or as a decimal. It could also be expressed as a percent (for example, a road with a rise of 10%). Make sure students are familiar with writing ratios as fractions or percents.

Planning Students may work on problems **7–9** in small groups. Students may work on problem **10** individually. This problem may also be used as an informal assessment.

Comments about the Problems

7. Problem **7** is critical because it is the first time students are asked to relate steepness and mathematical descriptions of steepness, such as the ratio of height to distance (slope) or the angle of elevation.

Students should discuss why height is measured vertically. Ask students, *How do you measure your height at home?* [vertically] *What would happen if you measured your height when you were standing at an angle and the measuring tape was held vertically?* [The measurement would be wrong.]

9. If students have difficulty seeing that a ratio increases, you might have them convert the ratio to a decimal or fraction.

10. Informal Assessment This problem assesses students' ability to measure angles and make scale drawings of situations involving steepness.

You may want to provide graph paper for students. Students might solve problems **d** and **e** using the same method used to solve **a, b,** and **c,** or they can solve these problems by observing the pattern that is being developed.

11. Copy the following table and fill it in using your data from problem **10.** Arrange your entries so that the angle measures increase from left to right.

Steepness Table

α (angle measure in degrees)					
h:d (ratio of height to distance)					

12. Use the table from problem **11** to make a graph of the height-to-distance ratio for a ladder leaning against a wall. Label your graph as shown below.

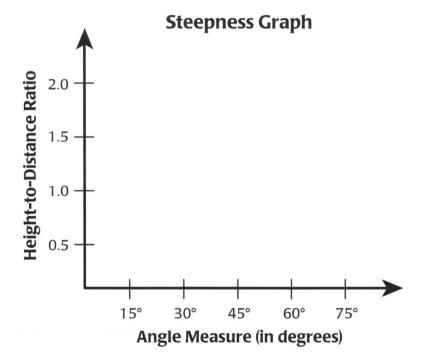

13. Explain the information shown in your graph. Compare your graph to your answer to problem **9.**

Suppose that it is safe to be on a ladder when the ratio *h:d* is larger than two and smaller than three.

14. Give a range of angles at which a ladder can be positioned safely.

11.

α (angle measure in degrees)	27°	30°
h:d (ratio of height to distance)	1:2 = 0.5	1:1.7 = 0.57

45°	60°	63°
1:1 = 1.0	1.7:1 = 1.7	2:1 = 2.0

12.

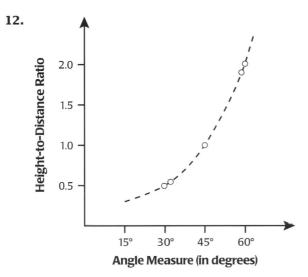

13. The larger the angle, the larger the ratio of height to distance. The ratio increases faster than the angle increases.

14. To be safe, the ladder can be positioned at any angle between 64° and 71°.

Materials rulers (one per student); ladder, optional (one per class); graph paper, optional (two sheets per student)

Overview Students make a table and a graph of steepness and angles of elevation. They describe the relationship between the two. They also investigate the height-to-distance ratios at which a ladder can be positioned safely.

About the Mathematics The graph students construct for problem **12** will be used again, in Section D, where students investigate the graph of the tangent in the context of glide ratios.

Planning Students may work on problems **11–13** individually. Problems **11** and **12** may also be assigned as homework. Be sure to discuss problem **13** in class. Students may continue working on problem **14** in small groups.

Comments about the Problems

11–12. Homework These problems may be assigned as homework.

11. This problem is critical because it focuses students' attention on the relationship between steepness and angle of elevation. Make sure students have the correct data from problem **10**. The ratio *h:d* may be expressed as a fraction or a decimal.

12. Again, students need the correct answers from problem **10**. The height-to-distance ratio should now be expressed as a decimal because this makes it easier to label the axes.

14. A height-to-distance ratio between two and three is safe for most ladders, but it depends on the ladder.

Summary

As the angle between a ladder and the ground increases, the height of the wall that can be reached by the top of the ladder increases. At the same time, the distance between the foot of the ladder and the wall decreases.

In the same way, as the angle between a ray of sunlight and the ground increases, a shadow on the ground becomes shorter.

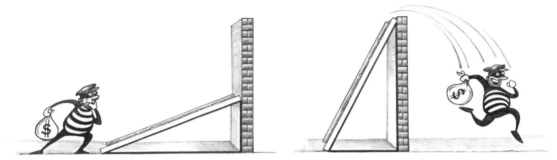

The steepness of a ladder can be measured in the following two ways:

- by the angle (the larger the angle, the steeper the ladder),

- by the ratio of the height to the distance, or *h:d* (the larger the ratio, the steeper the ladder).

Summary Questions

Recall the model of the canyon from Section A. The drawing on the right is a cross-section of another canyon model. The numbers indicate the scale of the height and the width of the ledges and the width of the river.

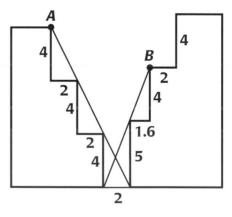

15. Which vision line is steeper: the one from point *A* down to the river or the one from point *B* down to the river? Support your answer with information about the angle between the vision line and the river, and the ratio of the height to the distance.

15. Vision line B is steeper. Sample response:

Vision line A is at an angle of 63°. The height is 12, and the distance is 6. This means that the ratio of height to distance is 12:6, or 2:1. We already found that a ratio of 2:1 gives an angle of 63°.

The height from point B to the river is 9, and the distance is 3.6, which makes a ratio of 9:3.6 or 2.5:1. Since 2.5:1 is larger than the ratio of vision line A, the angle must be larger, making for a steeper vision line.

Materials rulers (one per student); protractors or compass cards (one per student)

Overview Students read the Summary, which reviews the main concepts covered in this section. Then they determine which of two vision lines in a canyon is steeper, and support their answer using the height-to-distance ratio and the angle formed at the base.

About the Mathematics This section makes a connection with Section A, where students investigated vision lines in the context of the Grand Canyon. You may refer to the activity with the tables on Student Book page 3. The vision line of a taller person (or a person sitting nearer the edge of the canyon wall) was steeper; therefore this person could see more of the river.

Planning Students may work on problem **15** individually. This problem may also be used as an informal assessment. Discuss students' answers in class. After students complete Section C, you may assign appropriate activities from the Try This! section, located on pages 47–50 of the *Looking at an Angle* Student Book, for homework.

Comments about the Problems

15. Informal Assessment This problem assesses students' ability to understand the concept of steepness, to measure angles, and to make relative comparisons involving steepness problems. It also assesses their ability to understand the relationship among steepness, angle, and height-to-distance ratio and to use ratios to solve problems involving steepness.

Students can determine the steeper vision line by comparing the ratios of the two vision lines, concluding which ratio is bigger, and deducing which angle must be larger. Students might also compare the sizes of the angles, and deduce which height-to-distance ratio must be bigger. Either approach is fine, as long as students use the data of ratio and angle to support their answers.

You may want to discuss the advantages and disadvantages of using ratios instead of protractors to find angles.

SECTION D. GLIDE ANGLES

Work Students Do

In this section, students investigate the flight paths of hang gliders. First, they consider flight paths that are equally steep but begin at different heights. Next, they compare flight paths that vary in steepness. Students consider the relationship between the launch angle, the launch height, and the distance covered along the ground. They learn about the glide ratio, study the relationship between launch height and distance traveled, and compare flight paths where the height has a value of one. Then students learn that the ratio of height to distance is also called the tangent of angle α. They find the tangents of angles in several right triangles, explore the relationship between the size of the tangent of the glide angle and the distance a glider will fly, and learn to use formal tangent notation.

Goals

Students will:

- understand the concept of glide ratio, or tangent;
- measure angles;*
- make scale drawings of situations involving steepness;
- make relative comparisons involving steepness problems;
- understand the relationship between steepness, angle, and height-to-distance ratio;*
- choose appropriate views (top, side, or front) to draw situations involving steepness;*
- understand the correspondences between contexts involving steepness that may be represented with a right triangle;
- use ratios to solve problems involving steepness;
- solve problems involving tangents.

These goals are assessed in other sections of the unit.

Pacing

- approximately four 45-minute class sessions

Vocabulary

- glide angle
- glide ratio
- tangent of an angle

About the Mathematics

In this section, the tangent ratio (the relationship between the lengths of the legs of a right triangle) is formally introduced. Students first consider the different height and distance relationships that exist for paths of a hang glider when the steepness of the path remains fixed. In essence, students are examining a set of similar triangles in which the lengths of the sides vary but the angles remain constant. They then investigate the glide ratio—a more general ratio in which the steepness of the path varies and the height is fixed at 1.

After the tangent ratio is introduced, students find the tangents of angles in different right triangles, given the lengths of the two legs. They use their knowledge of tangents to find unknown side length and angle measurements in right triangles. See the About the Mathematics section on page 81 of the Teacher Guide for more on the tangent ratio.

Materials

- Student Activity Sheet 18, page 134 of the Teacher Guide (one per student)
- model glider airplane, page 81 of the Teacher Guide, optional (one per class)
- centimeter rulers, pages 83, 85, 87, and 101 of the Teacher Guide (one per student)
- paper ($8\frac{1}{2}$" × 11"), page 85 of the Teacher Guide, optional (one sheet per student)
- tape measure, page 85 of the Teacher Guide, optional (one per class)
- graph paper, pages 87 and 91 of the Teacher Guide, optional (one sheet per student)
- protractors or compass cards, page 97 of the Teacher Guide (one per student)
- scientific calculators, pages 89, 91, 97, 99, 101, and 103 of the Teacher Guide, optional (one per student)
- compasses, page 103 of the Teacher Guide (one per student)

Planning Instruction

This section begins with a discussion of hang gliders. To introduce this section, you might want to demonstrate a model glider airplane taking off at various heights. You could have students measure the height and distance along the ground for various flight paths. They could draw the side view of a flight path in the shape of a right triangle.

You may want to have students work on problems 1 and 29 as whole-class activities. Students may work on problems 2, 3, 6, 7, 9, 10, 13, 14, 17–19, 21–24, 30, and 31 in small groups. The remaining problems may be done individually.

Problem 29 and the activity on page 37 of the Student Book are optional. If time is a concern, you may omit these problems or assign them as homework.

Homework

Problems 8 (page 86 of the Teacher Guide) and 25–28 (page 100 of the Teacher Guide) may be assigned for homework. The Extension (page 89 of the Teacher Guide) and the Writing Opportunities (pages 91 and 103 of the Teacher Guide) may also be assigned as homework. After students complete Section D, you may assign appropriate activities from the Try This! section, located on pages 47–50 of the *Looking at an Angle* Student Book. The Try This! activities reinforce the key mathematical concepts introduced in this section.

Planning Assessment

- Problem 4 can be used to informally assess students' ability to make scale drawings of situations involving steepness, make relative comparisons involving steepness problems, and use ratios to solve problems involving steepness.

- Problems 6, 16, and 30 can be used to informally assess students' ability to understand the concept of glide ratio, or tangent. Problem 6 also assesses their ability to make relative comparisons involving steepness problems.

- Problem 20 can be used to informally assess students' ability to understand the correspondences between contexts involving steepness that may be represented with a right triangle.

- Problems 25, 27–29, and 31 can be used to informally assess students' ability to solve problems involving tangents.

- Problem 26 can be used to informally assess students' ability to make relative comparisons involving steepness problems and understand the relationship between steepness, angle, and height-to-distance ratio.

D. GLIDE ANGLES

HANG GLIDERS

Hang gliders are light, kite-like gliders that carry a pilot in a harness. The pilot takes off from a hill or a cliff against the wind. The hang glider then slowly descends to the ground.

When pilots make their first flight with a new glider, they are very careful because they do not know how quickly the glider will descend.

Marianne, the pilot in the above picture, decides to make her first jump from a 10-meter-high cliff. She glides along a straight line, covering 40 meters of ground.

Materials model glider airplane, optional (one per class)

Overview Students begin this section by reading about hang gliders. There are no problems on this page for students to solve.

About the Mathematics The context of the hang glider is similar to the other three contexts featured in the previous sections. All contexts involve steepness and can be represented with a right triangle: the steepness of the vision line in the context of the canyon, the steepness of the vision line of the captain on a ship, the steepness of ladders, and, in this section, the steepness of the flight path of the hang glider.

Students investigate the glide ratio: a general ratio in which the steepness of the path varies and the height is fixed at 1. The steeper path will have a large glide ratio such as 1:1 or 1:0.5, with a value larger than 1. A less steep path would have a glide ratio of 1:2 or 1:20, with a value less than 1. The glide angle is also investigated. The glide angle (as shown in the figure below) is the angle the flight path makes with the ground. A steeper path has a glide angle of between 45 and 90 degrees. A less steep path has a glide angle of between 0 and 45 degrees. In reality, the safest glide angle for a hang glider is about 6 degrees or smaller.

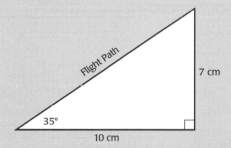

The right triangles of hang gliding are used to study the tangent function of the angle of any right triangle. The ratio of height to distance is equal to the tangent of the angle opposite the height.

In the example shown here, the tangent of the 35° angle is 0.7. By looking at a table of tangent values for angles ranging from 0 to 90 degrees (see Student Book page 44), it can be seen that the tangent ratio increases slowly for the first 45 degrees, ranging from 0 to 1, and then increases rapidly as the angle increases from 45 to 90 degrees. The section ends with students solving problems using the tangent relationship.

Shown below is a picture of Marianne's first flight path.

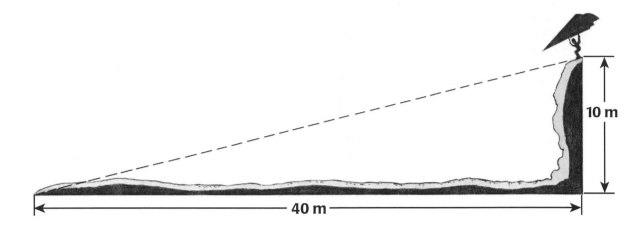

After several successful flights, she decides to go to a higher cliff. This cliff is 15 meters high.

1. How much ground distance does the glider cover from the higher cliff? (*Note:* Assume that the steepness of the flight path remains the same.)

2. Marianne makes flights from three cliffs that are 20 meters, 50 meters, and 100 meters high. How much ground distance does the glider cover on each flight?

Marianne has designed a glider that can travel farther than her first one. With the new glider, Marianne claims, "When jumping from a 10-meter-high cliff, I can cover 70 meters of ground!"

3. a. Draw a side view of Marianne's flight path with the new glider.

b. Copy the table below and complete it for the new glider.

Height (in meters)	10	25	100			
Distance (in meters)	70			245	1,000	

1. The glider covers 60 meters. Students may use a ratio table to solve this problem, as shown below.

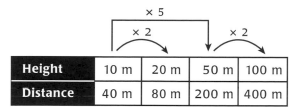

	÷ 2		× 3
Height	10 m	5 m	15 m
Distance	40 m	20 m	60 m

2. On the flight from the 20-meter cliff, the glider covers 80 meters. On the flight from the 50-meter cliff, the glider covers 200 meters. On the flight from the 100-meter cliff, the glider covers 400 meters. Again, students may use a ratio table to solve this problem.

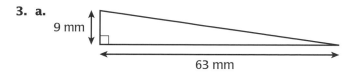

	× 5			× 2
Height	10 m	20 m	50 m	100 m
Distance	40 m	80 m	200 m	400 m

3. a.

9 mm

63 mm

b.

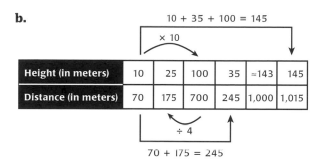

10 + 35 + 100 = 145

× 10

Height (in meters)	10	25	100	35	≈143	145
Distance (in meters)	70	175	700	245	1,000	1,015

÷ 4

70 + 175 = 245

Materials centimeter rulers (one per student)

Overview Students use ratio tables to find out how far a glider will fly from various heights when the distance for one height is given.

About the Mathematics At the start of the section, students look at a glider with a fixed height-to-distance ratio, so the steepness of its flight path is constant. Side-view drawings of different flights of this glider will look like similar triangles; the angle between the flight path and the ground remains the same.

Next, students investigate gliders with a different height-to-distance ratio. Further on in the section, students need to develop a measure (the "glide ratio") in order to compare the performances of different hang gliders.

Planning You may want to have students work on problem **1** as a whole-class activity. They may work on problems **2** and **3** in small groups. Discuss students' answers.

Comments about the Problems

1–3. For these problems, *distance* refers to the distance covered on the ground. So, when a problem asks how far the glider flew, the solution is not the length of the flight path of the glider (the hypotenuse of the right triangle formed by height, distance, and flight path). The solution is the ground distance covered by the glider.

1–2. If students have difficulty, you might have them sketch side views of the flight paths. Also, suggest that students use a ratio table to solve these problems.

3. a. Tell students to sketch a side-view that is drawn to scale. You might have students determine an appropriate scale themselves. For example, 1 centimeter could represent 10 meters.

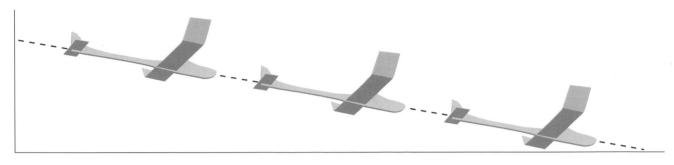

The above picture is based on three separate photographs, taken one after the other. It shows a model glider that is used in laboratory experiments. By taking three pictures within a short period of time, you can clearly see the path of the glider.

4. In your notebook, trace the path of the above glider and make a scale drawing, similar to the drawing on top of page 36, of its possible flight path to answer the following questions:

   a. If the glider in the picture is launched from a height of 5 meters, how far will the glider fly before landing?

   b. How far will the glider fly from a 10-meter cliff?

   c. Compare the distances covered by Marianne's two hang gliders and this model glider. If all three are launched from 10 meters, which one flies the farthest? Explain.

Activity

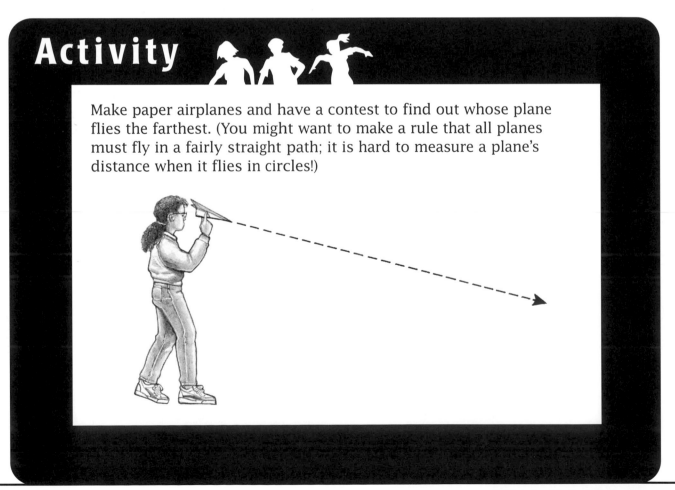

Make paper airplanes and have a contest to find out whose plane flies the farthest. (You might want to make a rule that all planes must fly in a fairly straight path; it is hard to measure a plane's distance when it flies in circles!)

4. a. about 27 meters. Answers may vary somewhat due to measurement errors. Strategies will vary. Sample strategy:

The ratio is 3:16.3, which is about 1:5.4. Launched from a height of 5 meters, the glider will travel about 27 meters.

3.0 cm

16.3 cm

b. Launched from a height of 10 meters, the plane will travel for about 54 meters.

c. If all three gliders are launched from a height of 10 meters, glider 1 will travel 40 meters, glider 2 will travel 70 meters, and the model glider will travel 54 meters. Glider 2 flies the farthest.

Materials centimeter rulers (one per student); paper ($8\frac{1}{2}$" × 11"), optional (one sheet per student); tape measure, optional (one per class)

Overview Students trace the flight path of a glider with a given steepness and determine how far the glider will fly if it is launched from a given height. They compare the performance of the glider with those of the gliders on page 36 of the Student Book. Students also build paper airplanes and have a contest to find out whose plane flies the farthest.

Planning Students may work on problem **4** individually. This problem may also be used as an informal assessment. Discuss students' answers in class.

The activity is optional. If time is a concern, you may omit the activity or have students design and test their paper planes at home. For the competition, ask students to think of a fair way to compare the performances of their planes. For example, students might decide to throw all planes from the same height and have the thrower stand behind a line drawn on the floor. Students should measure the distance each plane covers on the ground. You may want to have students graph the results of their competition.

Comments about the Problems

4. Informal Assessment This problem assesses students' ability to make scale drawings of situations involving steepness, to use ratios to solve problems involving steepness, and to make relative comparisons involving steepness problems.

a. When students trace the glider's path, make sure that they maintain the glider's steepness. Students can do this by aligning the notebook paper so that its edges are exactly lined up with the edges of the Student Book page, or so that its edges are parallel to the edges of the Student Book page. Students should draw the height and a line that represents the ground (and that intersects with the flight path). The height and the line that represents the ground should form a right angle.

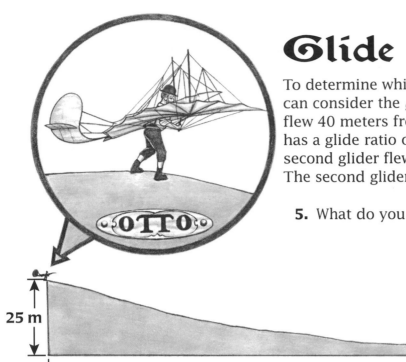

Glide Ratio

To determine which hang glider travels farther, you can consider the *glide ratio*. Marianne's first glider flew 40 meters from a 10-meter cliff. This glider has a glide ratio of 1:4 (one to four). Marianne's second glider flew 70 meters from a 10-meter cliff. The second glider has a glide ratio of 1:7.

5. What do you think a glide ratio is?

25 m

—— 185 m ——

Otto Lilienthal made more than 2,000 flights with hang gliders at the end of the 19th century. Suppose that on one of his flights from the Rhinower Hills near Berlin, Germany, he started from a height of 25 meters and covered 185 meters of ground distance as shown above. On his next flight, suppose he redesigned his glider a bit, started from a height of 20 meters, and traveled a distance of 155 meters along the ground.

6. What are the glide ratios of Otto's two gliders? Which glider can travel farther?

7. Suppose that a glider has a glide ratio of 1:8. It takes off from a cliff and covers 120 meters of ground distance. How high is the cliff?

8. Make scale drawings to represent the following glide ratios:

    a. 1:1   **b.** 1:2   **c.** 1:4

    d. 1:10   **e.** 1:20

5. Answers will vary. Students should note that a glide ratio is the ratio of height to distance. It tells you how far a glider can fly (in meters) if it is launched at a height of one meter.

6. 25:185 or 1:7.4 or about 1:7

20:155 or 1:7.75 or about 1:8

The second glider can travel farther because it has a slightly better glide ratio.

7. The cliff is 15 meters high. Students may use a ratio table.

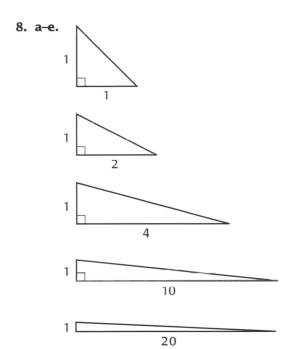

Height	1	10	5	15
Distance	8	80	40	120

80 + 40 = 120

8. a–e.

Materials centimeter rulers (one per student); graph paper, optional (one sheet per student)

Overview Students explain what a glide ratio is. They compare the performances of two gliders using glide ratios. Finally, they make scale drawings showing the flight paths of gliders with given glide ratios.

About the Mathematics The term *glide ratio* is used in aviation as a measure for the performance of gliders. It is defined as the height-to-distance ratio.

Planning Students may work on problem **5** individually. They may do problems **6** and **7** in small groups. Problem **6** can also be used as an informal assessment. Students may do problem **8** individually. This problem can also be assigned as homework.

Comments about the Problems

5. Encourage students to write extensively about what they think a glide ratio is. Discuss students' answers. Be sure that all students understand the definition of *glide ratio* before they continue.

6. Informal Assessment This problem assesses students' ability to understand the concept of glide ratio and make relative comparisons involving steepness problems.

Students will need to simplify the ratios in order to compare the planes. Students may remember the process of simplifying ratios from the work they did on page 34 of the Student Book.

7. Students may use a ratio table to solve this problem.

8. Homework This problem may be assigned as homework. You might want to have students use graph paper to make their scale drawings so they can align their right angles with the lines on the paper.

Did You Know? Otto Lilienthal was an important aviation pioneer. Orville and Wilbur Wright, the brothers who built the first powered airplane that was capable of sustained flight, drew on the work of Otto Lilienthal. Before trying to build powered airplanes, the Wright brothers built three glider airplanes.

In Section C when you studied ladders at different angles, you made a table similar to the one below showing the angle between the ladder and the ground, and the ratio of the height to the distance.

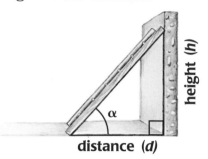

Ladder Steepness

α	27°	30°	45°	60°	63°
h:d	0.5	0.6	1	1.7	2

You can organize your information about the steepness of the glide path of a hang glider with a similar table. The angle that the hang glider makes with the ground as it descends is called a *glide angle*.

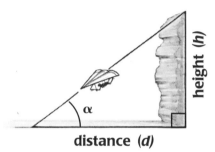

Glide Path Steepness

Glide Angle α					
Glide Ratio *h:d*	1:1	1:2	1:4	1:10	1:20

9. Copy the above table into your notebook. Fill in the missing glide angles by measuring the angles in each of the scale drawings you made for problem **8** with a protractor.

Glide ratios can also be expressed as fractions or decimals.

10. Which of the following glide ratios are equivalent?

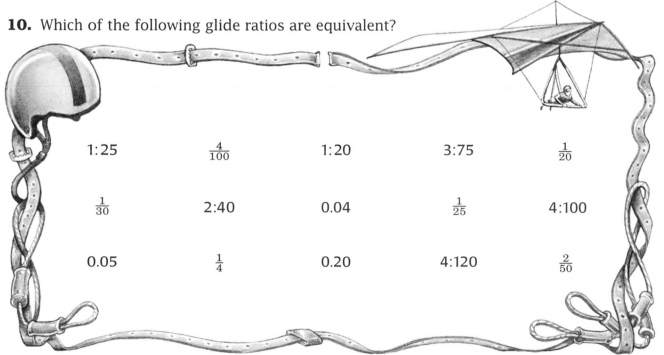

1:25 $\frac{4}{100}$ 1:20 3:75 $\frac{1}{20}$

$\frac{1}{30}$ 2:40 0.04 $\frac{1}{25}$ 4:100

0.05 $\frac{1}{4}$ 0.20 4:120 $\frac{2}{50}$

9.

Glide Path Steepness

Glide Angle α	45°	27°	14°	6°	3°
Glide Ratio *h:d*	1:1	1:2	1:4	1:10	1:20

10. The following glide ratios, decimals, and fractions are equivalent:

$1:25 = 3:75 = 4:100 = \frac{1}{25} = \frac{2}{50} = \frac{4}{100} = 0.04$

$1:20 = 2:40 = \frac{1}{20} = 0.05$

$4:120 = \frac{1}{30}$

(*Note:* Not all of the glide ratios listed in problem **10** are equivalent to another glide ratio.)

Materials scientific calculators, optional (one per student or group of students)

Overview Students make a table of the steepness of different flight paths using the information they gathered on page 38 of the Student Book. Also, they match equivalent glide ratios that are expressed as ratios, fractions, and decimals.

About the Mathematics Students learned about equivalent ratios and fractions and about the relationships among ratios, fractions, and decimals in the units *Fraction Times, More or Less,* and *Ratios and Rates.* Make sure students are familiar with expressing ratios as fractions or percents.

Planning Students may work on problems **9** and **10** in small groups. Have students compare their answers. You may want to discuss students' answers as a whole class.

Comments about the Problems

9. For this problem, students will need the scale drawings they made for problem **8** on page 38 of the Student Book. If students' scale drawings are incorrect, their angle measures may also be inaccurate.

10. Encourage students to solve this problem without using a calculator.

Extension Glide ratios can also be expressed as percents. For example, $\frac{1}{5}$ and 0.20 are equivalent to 20%; 4:100, $\frac{1}{25}$, and 0.04 are equivalent to 4%; 1:30 is equivalent to $3\frac{1}{3}$%; and 1:20 is equivalent to 5%. You may want to have students express the ratios, fractions, and decimals given in problem **10** as percents.

11. In your notebook, graph the information from the table in problem **9.** Label the axes as shown on the right, and take care to be precise when making a scale for the axes.

12. Explain what your graph shows.

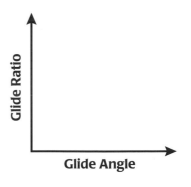

Suppose that it is safe to fly gliders that have a glide ratio smaller than 1:10.

13. What is the largest glide angle that is safe?

14. Suppose three gliders have the following glide ratios:

- Glider 1   1:27

- Glider 2   0.04

- Glider 3 $\frac{3}{78}$

Which glider is the safest? Explain.

11.

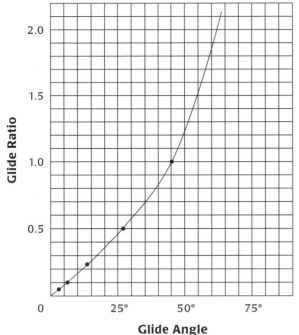

12. Explanations will vary. Sample explanations:

As the glide ratio gets larger, so does the angle.

The smaller the glide ratio, the smaller the angle.

13. The largest glide angle that is safe is about 6°.

14. Glider 1 has the smallest glide angle, so it can travel farther than the other two gliders. The glider that can travel the farthest is also the safest.

In order to compare the gliders, students may write their glide ratios as decimals:

Glider 1   1:27 = 0.037
Glider 2   0.04
Glider 3 $\frac{3}{78}$ = 0.038

Or, they may convert the decimal notations to glide ratios:

Glider 1   1:27
Glider 2   0.04 = 1:25
Glider 3 $\frac{3}{78}$ = 1:26

Materials graph paper, optional (one sheet per student); scientific calculators, optional (one per student or group of students)

Overview Students graph the data on flight path steepness that they collected on page 39 of the Student Book. They also compare the performances of gliders to determine which glider is safest.

Planning Students may work on problems **11** and **12** individually. Problems **13** and **14** may be done in small groups. Discuss students' answers in class.

Comments about the Problems

11–12. If students have difficulty, you might remind them of the graph they made of the steepness of ladders on page 33 of the Student Book. Ask students, *How did the steepness of the ladder affect the size of the angle between the foot of the ladder and the ground?* [The steeper the ladder, the larger the angle.] *How does the steepness of the flight path affect the size of the glide angle?* [The steeper the flight path, the larger the glide angle.]

13. Encourage students to solve this problem using the graph they made for problem **11.** Any angle smaller than six degrees is safe, since it means that the glider will travel a long distance. A glider flying at a very steep angle is more likely to crash.

14. Students should realize that they will have to convert all the glide ratios to decimals or ratios in order to compare them.

Writing Opportunity You might ask students to write about the similarities they see between the situations involving ladders and gliders. In their descriptions, students should use the terms *steepness, height-to-distance ratio,* and *angle.*

From Glide Ratio to Tangent

The relationship between the glide ratio and the glide angle is very important in hang gliding as well as in other applications, such as the placement of a ladder. For this reason, there are several ways to express this ratio and angle.

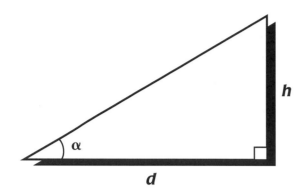

glide ratio = h:d

glide angle = α

The ratio h:d or $\frac{h}{d}$ is also called the *tangent of angle* α or:

$$\tan \alpha = \frac{h}{d}$$

For a glide ratio of 1:1, the glide angle is 45°, so:

$$\tan 45° = \frac{1}{1} = 1$$

Suppose that another one of Otto's hang gliders, shown above, has a glide ratio of 1:7. This means that the tangent of the glide angle is 1 to 7 (or $\frac{1}{7}$).

Overview Students study a formal description of the concept of tangent. They are also introduced to tangent notation. There are no problems on this page for students to solve.

About the Mathematics At this point, what was known to students as "steepness" in the context of ladders or "glide ratio" in the context of gliders is formally named *tangent.* The tangent is the ratio of vertical height to horizontal distance in any context that can be represented with a right triangle.

Planning You may want to read and discuss this page together with students. Encourage students to ask questions and discuss anything that may be unclear to them. Students will use formal tangent notation throughout the rest of this section.

A Matter of Notation

Suppose that a glider follows the flight path shown on the right.

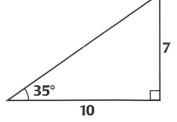

From this information, you can say that the glide ratio of a 35° angle is 0.7 (or 7:10). You can also write this information in the following way:

$\tan 35° = \frac{7}{10} = 0.7$

For the situation shown on the right, you can state the following:

$\tan A = \frac{25}{53} \approx 0.47$

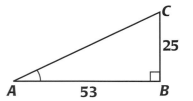

15. Complete the statements below for each of the following triangles:

a.

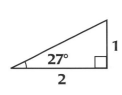

$\tan \underline{\ ?°\ } = \underline{\ ?\ }$

b.

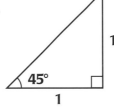

$\tan \underline{\ ?°\ } = \underline{\ ?\ }$

c.

$\tan \underline{\ ?°\ } = \underline{\ ?\ }$

d.

$\tan \underline{\ ?°\ } = \underline{\ ?\ }$

Peter wants to buy a balsa wood flyer for his nephew, but he is not sure which flyer to buy. The salesperson at the hobby store claims, "The smaller the tangent of the glide angle, the better the glider."

16. Is the salesperson correct? Explain.

15. **a.** tan 27° = 1:2, or $\frac{1}{2}$

   b. tan 45° = 1:1, or 1

   c. tan 63° = 2:1, or 2

   d. tan 72° = 3:1, or 3

16. Yes, the salesperson is correct. Explanations will vary. Sample explanation:

The smaller the tangent of the glide angle, the better the glide ratio. For instance, a glide ratio of 1:20 means that for every meter of launching height, the glider will cover 20 meters on the ground. A smaller glide ratio such as 1:200 means that the glider covers 200 meters on the ground if it is launched from a height of 1 meter.

Overview Students find the tangents of angles in four right triangles. They also investigate how the tangent of the glide angle affects the performance of a glider.

Planning Students should read the text at the top of the page carefully. Students may work on problems **15** and **16** individually. Discuss students' answers in class. Problem **16** may also be used as an informal assessment.

Comments about the Problems

15. This problem focuses on tangent notation. Students do not have to calculate or measure anything. They may refer to the example at the top of the page to see how statements like these are written. Note that the measurements shown in the solutions column are approximations. For example, tan 63° is closer to 1.9626 than to 2.

16. **Informal Assessment** This problem assesses students' ability to understand the concept of glide ratio, or tangent.

Students must realize that a glider with a small glide ratio or tangent will fly relatively far (considering the height).

Extension You might ask students to comment on the following statement: "The smaller the glide angle, the better the glider." Have students compare this statement to the one made by the salesperson on page 42 of the Student Book.

17. Suppose for triangle *ABC*, the measure of angle *B* is 90° and tan $A = \frac{3}{5}$.

    a. Make a scale drawing of triangle *ABC*.

    b. If you drew triangle *ABC* so that side *AB* measures 10 centimeters, what would be the length of side *BC*?

    c. What is the measure of angle *A* in triangle *ABC*?

The following table lists some angles and the approximate measures of their tangents.

Angle	0°	1°	2°	3°	4°	5°	31°	32°	33°	34°	35°
Tangent	0	0.02	0.04	0.05	0.07	0.09	0.60	0.63	0.65	0.68	0.70

Use the above table to answer the following problems:

18. **a.** Draw a side view of the flight path for a glider whose glide angle is 5°.

    b. What is the glide ratio for this glider?

19. If the glide angle is 35°, how much ground distance does a glider cover from a height of 100 meters?

20. If a ladder makes an 80° angle with the ground, what can you determine about the position of the ladder from the information that tan 80° equals about 5.7?

17. a.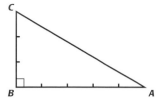

b. Side *BC* would measure 6 centimeters. A ratio table may be used, as shown below.

×2

Height *BC*	3 cm	6 cm
Distance *AB*	5 cm	10 cm

c. Angle *A* measures about 31°.

18. a.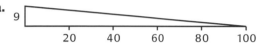

b. 9:100. The table tells you that for an angle of 5°, the tangent is 0.09, so the ratio must be 9:100. Therefore, when the glider is launched from a height of 9 meters, it covers a ground distance of 100 meters.

19. 143 meters. If the glide angle is 35°, the tangent is 0.70 or 7:10; when the glider takes off from a height of 7 meters, it covers 10 meters of ground distance. See ratio table below:

× 100   ÷ 7

Height (in m)	7	700	100
Distance (in m)	10	1,000	142.9

20. Answers will vary. Sample response:

The ladder reaches 5.7 meters up the wall if the bottom of the ladder is 1 meter away from the wall, as shown below.

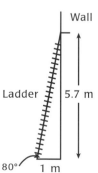

Materials centimeter rulers (one per student); protractors or compass cards (one per student); scientific calculators, optional (one per student)

Overview Students make scale drawings of right triangles. When the glide angle is given, students find the tangent, and vice versa. They use a table of angles and their corresponding tangents to solve problems.

Planning Students may work on problems **17–19** in small groups. Discuss their answers in class. Students may work on problem **20** individually, and it may be used as an informal assessment.

Comments about the Problems

17. a. There are several possibilities for making a scale drawing of the triangle. Any scale drawing is acceptable, as long as the ratio between height and distance is 3:5. Make sure students label vertices *A, B,* and *C* correctly.

b. If the distance is multiplied by two, the height must also be multiplied by two to maintain the same height-to-distance ratio.

c. No matter what size students' drawings are, they should find that the angle measures about 31°.

18. a. Students may first draw the 5° angle and then complete the triangle. The length of the distance they choose is not important, but it will be easier to find the height-to-distance ratio if students choose a convenient length.

b. Students can find the glide ratio by measuring the height and distance of the triangle they drew. But it is easier to use the table given on page 43 of the Student Book.

20. Informal Assessment This problem assesses students' ability to understand the contexts involving steepness that may be represented with a right triangle. Be sure students understand that the tangent ratio relates to triangles formed in many contexts, such as gliding, ladders leaning against walls, vision lines on boats, and so forth.

Angle (in degrees)	Tangent
0	0.000
1	0.017
2	0.035
3	0.052
4	0.070
5	0.087
6	0.105
7	0.123
8	0.141
9	0.158
10	0.176
11	0.194
12	0.213
13	0.231
14	0.249
15	0.268
16	0.287
17	0.306
18	0.325
19	0.344
20	0.364
21	0.384
22	0.404
23	0.424
24	0.445
25	0.466
26	0.488
27	0.510
28	0.532
29	0.554
30	0.577
31	0.601
32	0.625
33	0.649
34	0.675
35	0.700
36	0.727
37	0.754
38	0.781
39	0.810
40	0.839
41	0.869
42	0.900
43	0.933
44	0.966
45	1

Angle (in degrees)	Tangent
45	1
46	1.036
47	1.072
48	1.111
49	1.150
50	1.192
51	1.235
52	1.280
53	1.327
54	1.376
55	1.428
56	1.483
57	1.540
58	1.600
59	1.664
60	1.732
61	1.804
62	1.881
63	1.963
64	2.050
65	2.145
66	2.246
67	2.356
68	2.475
69	2.605
70	2.747
71	2.904
72	3.078
73	3.277
74	3.487
75	3.732
76	4.011
77	4.331
78	4.705
79	5.145
80	5.671
81	6.314
82	7.115
83	8.144
84	9.154
85	11.430
86	14.301
87	19.081
88	28.636
89	57.290
90	

The tables on the left and right show the relationship between the size of an angle and its tangent. You can also use a scientific calculator to find the tangent of an angle. Since calculators differ, you may want to investigate how to use the tangent key on your calculator. You can use the tables on this page to verify your work.

You can also use a scientific calculator to find angle measures if you know the tangent ratio.

Use either the tables or the tangent key on your scientific calculator to answer the following problems:

21. What do you know about a glider with a glide angle of 4°? a glide angle of 35°?

22. Explain why tan 45° = 1.

23. Which angle has a tangent of 2? of 3? of 4?

24. How much does the measure of the angle change when the tangent value changes:

  a. from 0 to 1?

  b. from 1 to 2?

  c. from 2 to 3?

  d. from 3 to 4?

  e. from 4 to 5?

21. Answers will vary. Sample responses:

A glider with a glide angle of 4° has a tangent of 0.070, or a glide ratio of 7:100. This glider can travel 100 meters when it is launched from a height of 7 meters.

A glider with a glide angle of 35° has a tangent of 0.700, or a glide ratio of 7:10. This glider is dangerous. If it is launched from a height of 7 meters, it will fly only 10 meters.

22. The tangent of 45° is equal to 1 because the height and distance in a right triangle with two 45° angles are equal. The ratio is 1:1, so the tangent is $\frac{1}{1}$, or 1.

23. An angle of about 63° or 64° has a tangent of 2.

An angle of about 71° or 72° has a tangent of 3.

An angle of about 76° has a tangent of 4.

24. a. The measure of the angle changes by 45°. When the tangent value is 0, the angle measures 0°; when the tangent value is 1, the angle measures 45°. The difference is 45°.

b. The measure of the angle changes by 18°. When the tangent value is 1, the angle measures 45°; when the tangent value is 2, the angle measures about 63°. The difference is 18°.

c. The measure of the angle changes by 8°. When the tangent value is 2, the angle measures about 63°; when the tangent value is 3, the angle measures about 71°. The difference is 8°.

d. The measure of the angle changes by 5°. When the tangent value is 3, the angle measures about 71°; when the tangent value is 4, the angle measures 76°. The difference is 5°.

e. The measure of the angle changes by 3°. When the tangent value is 4, the angle measures 76°; when the tangent value is 5, the angle measures about 79°. The difference is 3°.

Materials scientific calculators, optional (one per student)

Overview Students find angle measures for tangent values and vice versa, by using a scientific calculator or a tangent table.

Planning Introduce the tangent table or the tangent key on a scientific calculator in class. Make sure all students understand how to use one or both. Students will need to use the tangent table on Student Book page 44 or a scientific calculator to solve the remaining problems in this section. Students may work on problems **21–24** in small groups. Discuss their answers in class.

Comments about the Problems

21. Students may find the corresponding glide ratios in the tangent table (or by using a calculator) and reason from there.

22. This is an important tangent value to remember; you might call it a benchmark tangent value. You may ask students what they know about the tangent values for angles smaller than 45° and larger than 45°.

23–24. The purpose of these problems is to familiarize students with the use of the tangent table or the tangent key on a scientific calculator.

Extension If possible, obtain the software package *Gliding,* which is available from Sunburst, Pleasantville, New York. Students may use *Gliding* to get a feel for glide angles and their corresponding glide ratios. Students can play the game individually or as a competition between two students.

Solve each of the following problems using tangent ratios:

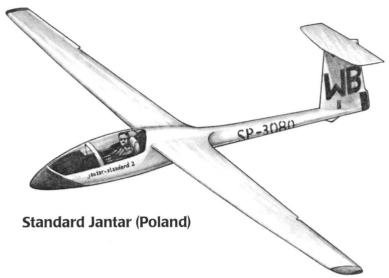

25. Suppose the glider on the right has a glide ratio of 1:40. It is flying over a village at an altitude of 230 meters, and it is 9 kilometers from an airstrip. Can it reach the airstrip? Explain.

Standard Jantar (Poland)

26. One glider has a glide ratio of 1:40, while another has a glide angle of 3°. Which glider flies farther? Explain why.

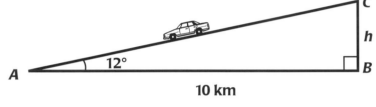

27. Compute the height of the road at point *C* for the drawing on the right.

28. At a distance of 160 meters from a tower, you look up at an angle of 23° and see the top of the tower. What is the height of the tower?

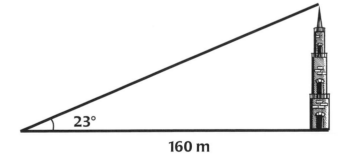

29. An electricity line pole makes an angle of 75° with the road surface, as shown on the right. How much does the road rise over a horizontal distance of 100 meters?

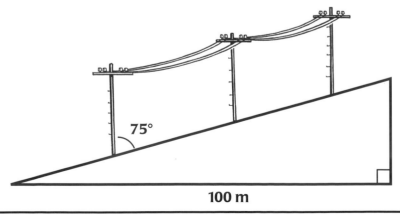

25. Yes, the plane can reach the airstrip. It can fly 40 meters for every meter in altitude that it has; $230 \times 40 = 9,200$ meters. So, it can reach the airstrip 9 kilometers, or 9,000 meters, away.

26. The glider with the glide ratio of 1:40 will fly farther. Strategies may vary. Sample strategy:

$\tan 3° = 0.052 \approx \frac{5}{100} = \frac{1}{20}$

$\frac{1}{40} = 0.025 = \tan 1.4°$

27. 2.1 kilometers. Strategies may vary. Sample strategy:

Tan 12° is about 0.21 or $\frac{21}{100}$, so the height is 2.1 kilometers or 2,100 meters. The following ratio table may be used:

÷ 10

Height (in m)	21	2.1
Distance (in m)	100	10

28. 67.2 meters. Strategies may vary. Sample strategy:

Tan 23° is about 0.42 or $\frac{42}{100}$, so the height of the tower is about 67.2 meters. The following ratio table may be used:

÷ 2   add

Height (in m)	42	21	63	4.2	67.2
Distance (in m)	100	50	150	10	160

× 3

29. 27 meters. Students should note that the left pole would make a 90° angle with the horizontal line. Therefore, as shown below, the angle formed by the road and the horizontal line is 15°, not 75°. Tan 15° is about 0.27 or about $\frac{27}{100}$.

So, $h/100 = \frac{27}{100}$, and $h = 27$ meters.

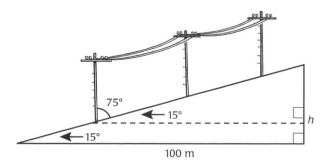

Materials centimeter rulers (one per student); scientific calculators, optional (one per student)

Overview Students solve problems involving tangents.

Planning Students may work on problems **25–28** individually. These problems may also be assigned as homework. Discuss students' answers. Students should work on problem **29** as a whole class. Problem **29** is optional. If time is a concern, you may omit this problem or assign it as homework. Problems **25–29** may be used as informal assessments.

Comments about the Problems

25. Informal Assessment This problem assesses students' ability to solve problems involving tangents.

If students have difficulty, you might remind them that the ratio of height-to-distance is 1:40. Ask students, *If the launch height is 230 meters, how far will the glider fly?* [9,200 meters] Remind students that there are 1,000 meters in a kilometer.

26. Informal Assessment This problem assesses students' ability to make relative comparisons involving steepness problems and their ability to understand the relationship between steepness, angle, and height-to-distance ratio.

27–29. Informal Assessment These problems assess students' ability to solve problems involving tangents.

29. This problem is difficult because the information necessary to solve it is given in an indirect way. Students must add something to the drawing in order to find the glide angle. Encourage students to sketch the situation.

Summary

The steepness of a ladder, the angle of the sun's rays, and the flight path of a hang glider can all be modeled by a triangle such as the one shown below.

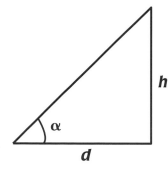

Steepness can be measured as the angle α or as the ratio $h{:}d$.

The ratio $h{:}d$ is also called the tangent of angle α, or $\tan \alpha = \frac{h}{d}$.

Summary Questions

30. On a calm day, a glider pilot wants to make a flight that covers 120 kilometers. The glider has a glide ratio of 1:40. From what height must the glider be launched?

31. A glider with a glide ratio of 1:28 is launched after being pulled by an airplane to 1,200 meters above Lake Havasu City in Arizona. Indicate on the map on **Student Activity Sheet 18** how far the glider can fly if there is no wind.

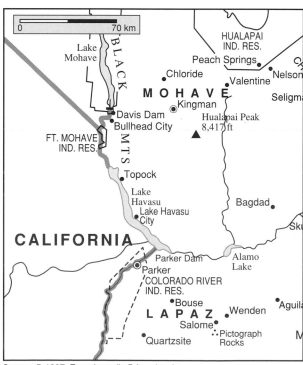

Source: © 1997, Encyclopædia Britannica, Inc.

30. The glider must be launched from a height of 3 kilometers. Students can use the following ratio table to solve the problem:

× 3

Height	1	3 km
Distance	40	120 km

× 3

31. 33,600 meters or 33.6 kilometers. Students may use the following ratio table to solve the problem:

200 + 1,000 = 1,200

× 10   × 2   × 10   × 5

Height	1	10	20	200	1,000	1,200
Distance	28	280	560	5,600	28,000	33,600

Students should draw a circle on the map to indicate all possible landing locations for the glider. The circle should be centered on Lake Havasu City.

Source: © 1997, Encyclopædia Britannica, Inc.

Materials Student Activity Sheet 18 (one per student); scientific calculators, optional (one per student); compasses (one per student)

Overview Students read the Summary, which reviews the main concepts covered in this section. They also use glide ratios to solve problems involving gliding.

About the Mathematics The Summary of this section outlines the main mathematical concepts of the entire unit. Each context that was featured in this unit may be represented with a right triangle: steepness of the vision line, steepness of the sun's rays, steepness of the ladder, and steepness of the glide path. Each serves to establish understanding of the tangent concept. Students may always remember the tangent as the "glide ratio" of a glider.

Planning Students may work on problems **30** and **31** in small groups. These problems may also be used as informal assessments. Discuss students' answers in class. After students complete Section D, you may assign appropriate activities from the Try This! section, located on pages 47–50 of the *Looking at an Angle* Student Book, for homework.

Comments about the Problems

30. Informal Assessment This problem assesses students' ability to understand the concept of glide ratio, or tangent. Encourage students to use a ratio table to solve this problem.

31. Informal Assessment This problem assesses students' ability to solve problems involving tangents. If students draw only one location for the glider to land, ask them whether there is another possible landing location, and repeat this question until they realize that they can indicate all possible landing locations by drawing a circle.

Writing Opportunity You may want to ask students to write about the connections they see between the four contexts from the unit: the vision lines, the light rays, the ladders, and the flight paths. Encourage students to describe these connections in as much detail as possible.

Assessment Overview

Students work on 11 assessment problems that you can use to collect additional information about what each student knows about vision lines, blind spots, angles, shadows, steepness, glide ratios, and tangents.

Goals

- understand the concepts of vision line, vision angle, and blind spot

- understand the concept of glide ratio, or tangent

- construct vision lines and blind spots (or light rays and shadows) in two- and three-dimensional representations

- measure blind spots (or shadows)

- measure angles

- understand the ratio between an object and its shadow caused by the sun for different times of the day and the year

- understand the relationship among steepness, angle, and height-to-distance ratio

- choose appropriate views (top, side, or front) to draw situations involving steepness

- use ratios to solve problems involving steepness

- solve problems involving tangents

Assessment Opportunities

Problems 1, 2, and 3

Problems 10 and 11

Problems 3, 6, and 8

Problems 4 and 5

Problem 9

Problems 6, 7, and 8

Problem 11

Problem 11

Problem 11

Problems 10 and 11

Pacing

When combined, the 11 assessment problems of the Investigating the Mammoth Rocks assessment will take approximately one or two 45-minute class sessions. For more information on how to use the 11 problems, see the Planning Assessment section on the next page.

About the Mathematics

The 11 assessment problems evaluate the major goals of the *Looking at an Angle* unit. Refer to the Goals and Assessment Opportunities sections on the previous page for information regarding the specific goals assessed in each assessment problem. Students may use different strategies to solve each problem. Their choice of strategies may indicate their level of comprehension of the problem. Consider how well students' strategies address the problem, as well as how successful students are at applying their strategies in the problem-solving process.

Materials

- Investigating the Mammoth Rocks assessment, pages 135–138 of the Teacher Guide (one of each per student)
- rulers or straightedges, pages 109, 111, and 113 of the Teacher Guide (one per student)
- compass cards or protractors, page 113 of the Teacher Guide (one per student)
- copies of the tangent table from page 44 of the Student Book, or scientific calculators with tangent keys, page 113 of the Teacher Guide (one per student)

Planning Assessment

You may want students to work on these assessment problems individually so that you can evaluate each student's understanding and abilities. Make sure that you allow enough time for students to complete the assessment problems. Students are free to solve each problem in their own way. They may choose to use any of the models introduced and developed in this unit to solve problems that do not call for a specific model.

Scoring

Answers are provided for all assessment problems. The method of scoring the problems depends on the types of questions in each assessment. Most questions require students to explain their reasoning or justify their answers. For these questions, the reasoning used by the students in solving the problems as well as the correctness of the answers should be considered as part of your grading scheme. A holistic scoring scheme can be used to evaluate an entire task. For example, after reviewing a student's work, you may assign a key word such as *emerging, developing, accomplishing,* or *exceeding* to describe his or her mathematical problem solving, reasoning, and communication.

On other tasks, it may be more appropriate to assign point values for each response. Students' progress toward the goals of the unit should also be considered. Descriptive statements that include details of a student's solution to an assessment activity can be recorded. These statements could provide insight into a student's progress toward a specific goal of the unit. Descriptive statements are often more informative than recording only a score and can be used to document students' growth in mathematics over time.

Use additional paper as needed.

Two scientists, Jorge and Theresa, are studying in the southwestern part of the United States. They are investigating rock formations, deserts, plants, and animals. They are an adventurous pair—hiking in canyons, climbing rocks, and hang gliding are some of their favorite ways to explore and investigate.

Today they are working in Mammoth Rock Country.

Mammoth Rock Country

Shown below is a map of the area where they are working.

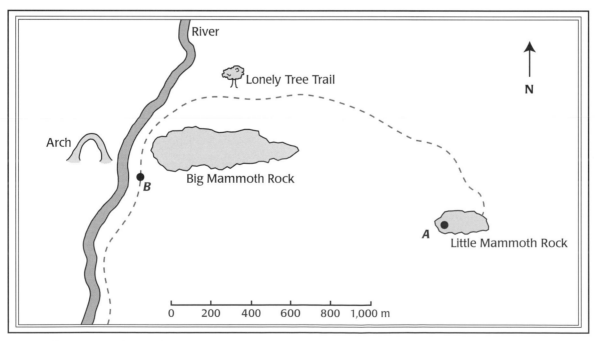

Map of Mammoth Rock Country

Overview Students read about the setting of this assessment, Mammoth Rock Country. They study a picture and a map of the area that will be used throughout this assessment. There are no problems on this page for students to solve.

Planning Have students read this page carefully and study the picture and the map.

Use additional paper as needed.

Shown below is a picture that was taken in Mammoth Rock Country.

1. Find a location on the map where the photographer may have been standing while taking this picture. Mark the spot with an "✕" on the map on the previous page.

Jorge and Theresa have just climbed Little Mammoth Rock. Theresa is at point *A* and about to take a hang glider flight. However, she will only start the flight when Jorge has descended the rock and has arrived at point *B*. Jorge is following the trail as he hikes. Theresa watches him from where she is sitting at point *A*. After a while, near Lonely Tree, Jorge disappears behind Big Mammoth Rock. Theresa waits and waits, but Jorge seems to stay out of sight forever. He has been in the "blind area" for half an hour.

2. Outline the blind area behind the rock on the map on the previous page.

3. How long is the part of the trail in that blind area? Make an estimate using the scale line.

Under normal circumstances, a person can walk about 2 or 3 kilometers per hour.

4. Is there a reason for Theresa to be worried? Explain your answer.

1. Answers will vary. Sample response:

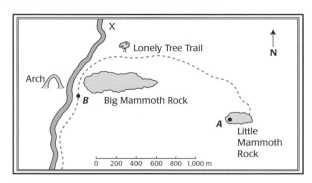

2.

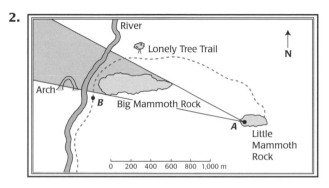

3. Answers will vary. Accept estimates in the range of 500–600 meters.

4. Answers and explanations will vary. Sample explanation:

Walking at 2 kilometers an hour means that a person can walk 1 kilometer in half an hour. Since that section of trail is closer to $\frac{1}{2}$ kilometer long than to 1 kilometer, Theresa should begin to worry.

Materials rulers or straightedges (one per student)

Overview Students look at a picture of Mammoth Rock Country and indicate where the photographer could have been when the picture was taken. They outline the blind spot behind a rock and estimate the length of the part of the trail that is in the blind spot.

Planning You may want students to work on these assessment problems individually.

Comments about the Problems

1. If students are unsure of their answer, you might suggest that they imagine standing in the location they have marked. Ask students, *What do you see to the right? What do you see to the left? What landmark do you not see? What do you see that is in front of something else?*

2. To solve this problem, students must remember that a blind area is the area that you cannot see from a certain position. Theresa cannot see through Big Mammoth Rock; she can only look past it to the left and right.

3. Students may use a straightedge or a ruler to measure the part of the trail that is in the blind spot. They can then find an answer using the map's scale.

4. To solve this problem, students will need the answer they found in problem **3**. If students' answers to problem **3** were wrong, but their solutions to problem **4** are otherwise correct, they should receive full credit.

Use additional paper as needed.

As Theresa waits for Jorge to reappear, she wonders if he is walking in the shadow of the rock or in the sun. She notices that the shadow of Little Mammoth Rock just reaches the bottom of Big Mammoth Rock. Theresa knows that Little Mammoth Rock is twice the height of Big Mammoth Rock. She draws the shadow on the map. It looks like this:

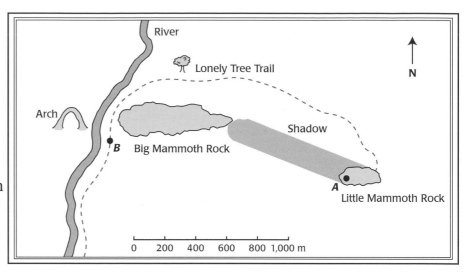

5. Draw the shadow of Big Mammoth Rock in the picture above. Be precise in drawing the shadow's length and direction.

Below you see three top-view drawings of the arch and its shadow at different times on one day.

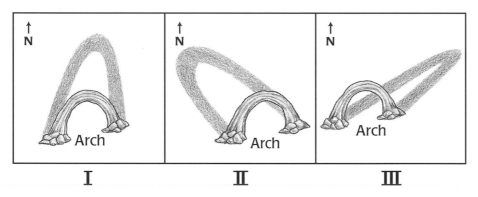

I   II   III

6. In which order were these drawings made?

7. Draw the shadow of the arch at sunset in the picture below. Be precise in drawing the shadow's length and direction.

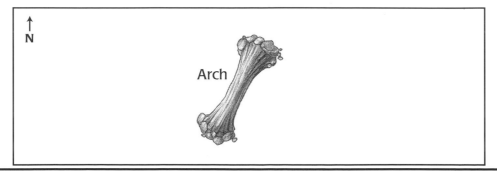

5.

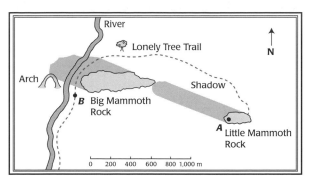

Since Little Mammoth Rock is twice as tall as Big Mammoth Rock, its shadow must be twice as long. Little Mammoth Rock's shadow measures about 3.2 centimeters on the student page, so Big Mammoth Rock's shadow should be about 1.6 centimeters long. The shadow of Big Mammoth Rock should point in the same direction as the shadow of Little Mammoth Rock.

6. The correct order of the pictures is II, I, III. Explanations will vary. Sample explanation:

In picture I, the sun is shining from the south; in picture II, it is shining from the southeast; and in picture III, it is shining from the southwest. Since the sun rises in the east and sets in the west, the correct order of the pictures is II, I, III.

7.

The sun sets in the west. At sunset, the shadow of the arch will be very long and will point to the east.

Materials rulers or straightedges (one per student)

Overview Students draw the shadow of one rock when the shadow of another rock is given. Students also look at drawings of the arch and its shadow, and determine the order of the drawings. Then they draw the shadow of the arch at sunset.

Planning You may want students to work on these assessment problems individually.

Comments about the Problems

5. Students should realize that the ratio between the height of an object and the length of its shadow is constant at a certain time of the day. For example, if the shadow of an object is three times the height of the object, then all shadows at that time of the day are three times as long as the height of the objects. At the same time of the day, an object that is twice as tall as another object must have a shadow that is twice as long.

6. You may want to remind students that the sun rises in the east and sets in the west. Encourage students to explain in writing why they put the drawings in a particular order.

Use additional paper as needed.

Below is a sketch of a side view of the two rocks. The line represents the sun's ray. The distance between the two rocks is 910 meters.

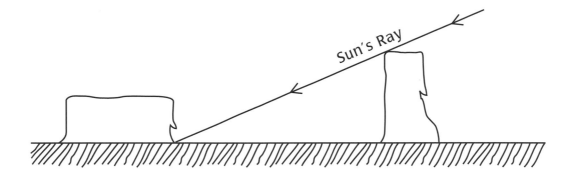

8. Measure the angle between the sun's ray and the horizon with a compass card or protractor.

9. Compute the height of Little Mammoth Rock. Show your work.

Hours later, the rock's shadow is six times longer than the height of the rock.

10. Compute the angle between the sun's ray and the horizon in this situation. Write an explanation of how you found the answer. Show your work.

Finally, Jorge reappears behind Big Mammoth Rock. He uses his walkie-talkie to tell Theresa that an injured hiker has been calling for help. The hiker is on the other side of the river near the arch (see the map). He asks Theresa to fly in her hang glider across the river to the hiker in order to help him. Theresa tells Jorge that she is not sure she can reach that spot. Her hang glider has a glide ratio of 1:6.5.

11. Can Theresa reach the wounded hiker with her hang glider from Little Mammoth Rock? Show the mathematics that supports your answer.

8. The angle is 23°.

9. 386 meters

The angle of the sun's ray is 23°.
The tangent of 23° is equal to the height of the rock divided by the length of its shadow:
$$\tan \text{ of } 23° = 0.424 = h/910 \text{ m}$$
$$h = 0.424 \times 910 \text{ m} = 386 \text{ meters}$$

10. The angle between the sun's ray and the ground is about 9.5°.

Some students may reason that since the shadow is six times longer than the height of the rock, the ratio of the height to the length of the shadow is $h:d = 1:6$, or 0.167. Using the table on Student Book page 44, students may see that a tangent of 0.167 corresponds to an angle of between 9° and 10°.

11. Yes. Strategies will vary. Sample strategy:

From Little Mammoth Rock to the arch is just under 2,000 meters. Theresa may use a ratio table to find out whether she can reach the injured hiker.

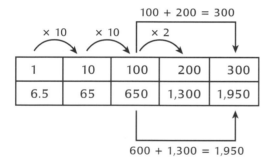

$$100 + 200 = 300$$

1	10	100	200	300
6.5	65	650	1,300	1,950

$$600 + 1,300 = 1,950$$

The ratio table shows that Theresa needs to launch her glider from a height of 300 meters. Since Little Mammoth Rock is 386 meters tall, she will make it.

Materials rulers or straightedges (one per student); compass cards or protractors (one per student); copies of the tangent table from page 44 of the Student Book, or scientific calculators with tangent buttons (one per student)

Overview Students measure the angle between the sun's ray and the horizon. They calculate the height of Little Mammoth Rock. Using the height-to-distance ratio, they also calculate the angle that will exist between the sun's ray and the horizon hours later. Finally, they determine how far a glider can travel given its glide ratio and the launching height.

Planning You may want students to work on these assessment problems individually.

Comments about the problems

9. To solve this problem, students will need the answers they found in problem **8.** If students' answers to problem **8** were wrong, but their solutions to problem **9** are otherwise correct, they should receive full credit.

10. Students should realize that they do not need to know which rock is meant; they can use the height-to-distance ratio to find the angle measure in the tangent table. If students have difficulty, you might encourage them to sketch the situation.

11. Students should use the glide ratio to calculate the maximum distance the glider can cover. Of course, the glider could also land at a shorter distance. To solve this problem, students will need the answers they found in problem **9.** If students' answers to problem **9** were wrong, but their solutions to problem **11** are otherwise correct, they should receive full credit.

Students can also find the distance that needs to be covered by using the scale on the map and determining the minimum glide ratio the glider needs to reach the hiker, which is 386:1,850, or 1:4.8. Theresa's glider has a glide ratio of 1:6.5, which is higher than the minimum ratio Theresa needs. So, she will be able to cover the distance.

Looking at an Angle
Glossary

The Glossary defines all vocabulary words listed on the Section Opener pages. It includes the mathematical terms that may be new to students, as well as words having to do with the contexts introduced in the unit. *(Note:* The Student Book has no glossary in order to allow students to construct their own definitions, based on their personal experiences with the unit activities.)

The definitions below are specific to the use of the terms in this unit. The page numbers given are from this Teacher Guide.

alpha (p. 72) the first letter of the Greek alphabet, used to represent an angle

beta (p. 72) the second letter of the Greek alphabet, used to represent an angle

blind area (p. 36) an area that cannot be seen because something is blocking the view; a blind spot

blind spot (p. 36) an area that cannot be seen because something is blocking the view; a blind area

gamma (p. 72) the third letter of the Greek alphabet, used to represent an angle

glide angle (p. 88) the angle a glider makes with the ground as it descends

glide ratio (p. 86) the ratio of how far a glider will drop in height (h) to the distance (d) it travels in a horizontal direction, or $h{:}d$

steepness (p. 68) the ratio of the height (h) to the distance (d), or $h{:}d$

tangent of an angle (p. 92) in a right triangle, the ratio of the side opposite the angle to the side adjacent to the angle, or height:distance

vision line (p. 12) an imaginary straight line from a person's eye to an object

BRITANNICA
Mathematics in Context

Blackline Masters

Dear Family,

Very soon your child will begin the *Mathematics in Context* unit *Looking at an Angle*. Below is a letter to your child that opens the unit, describing the unit and its goals.

In one section of the unit, your child will study how the angle of the sun or the distance of a light source affects the length of a shadow. You and your child might examine the changes in the length of a shadow of a tree and the corresponding angle of the sun at different times of the day. Your child then can model what you observed outside by holding a flashlight at different angles to change the length of the shadow of a household object.

You and your child might also talk about the blind spots that exist for the driver of a car. While your child moves to different positions around your parked car, you can sit in the driver's seat and indicate when you can or cannot see him or her. Your child then can sketch diagrams of your blind spots.

We hope you enjoy these ways of "looking at an angle" with your child.

Dear Student,

Welcome to *Looking at an Angle*.

In this unit you will learn about vision lines and blind areas. Have you ever been on one of the top floors of a tall office or apartment building? When you looked out the window, were you able to see the sidewalk directly below the building? If you could see the sidewalk, it was in your field of vision; if you could not see the sidewalk, it was in your blind spot.

The relationship between vision lines and light rays and the relationship between blind spots and shadows are some of the topics that you will explore in this unit. Have you ever noticed how the length of a shadow varies according to the time of day? As part of an activity, you will measure the length of the shadow of a stick and the corresponding angle of the sun at different times of the day. You will then determine how the angle of the sun affects the length of a shadow.

Besides looking at the angle of the sun, you will also study the angle that a ladder makes with the floor when it is leaning against a wall and the angle that a descending hang glider makes with the ground. You will learn two different ways to identify the steepness of an object: the angle the object makes with the ground and the tangent of that angle.

We hope you enjoy discovering the many ways of "looking at an angle."

Sincerely,

The Mathematics in Context Development Team

Sincerely,

The Mathematics in Context Development Team

© 1998 Encyclopædia Britannica Educational Corporation. This sheet may be reproduced for classroom use.

Use with *Looking at an Angle,*
page 4.

A

Use with *Looking at an Angle*, page 4.

B

Britannica Mathematics System

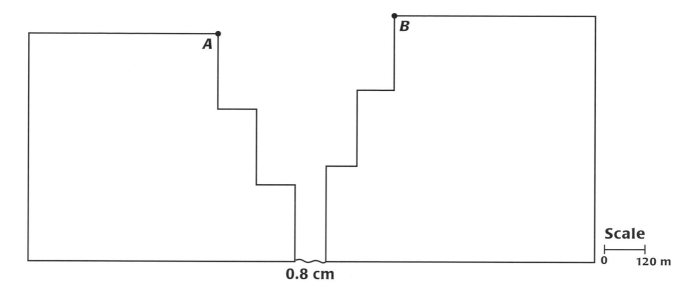

0.8 cm

Scale
0   120 m

8. Is it possible to see the river from point *A* on the left rim? Why or why not?

9. What is the actual height of the left canyon wall represented by the above scale model?

10. What is the actual width of the river represented by the above scale model?

11. If the river were 1.2 centimeters wide in the above scale model, could it be seen from point *A*? Explain.

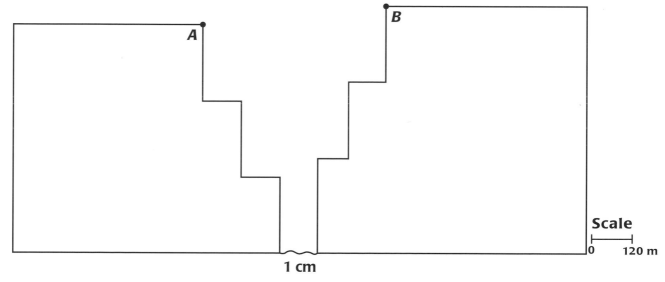

1 cm

Scale
0   120 m

12. In the above drawing, the river is 1 centimeter wide. Is it possible to see the river from point *B*? If not, which ledge is blocking your view? Explain.

© 1998 Encyclopædia Britannica Educational Corporation. This sheet may be reproduced for classroom use.

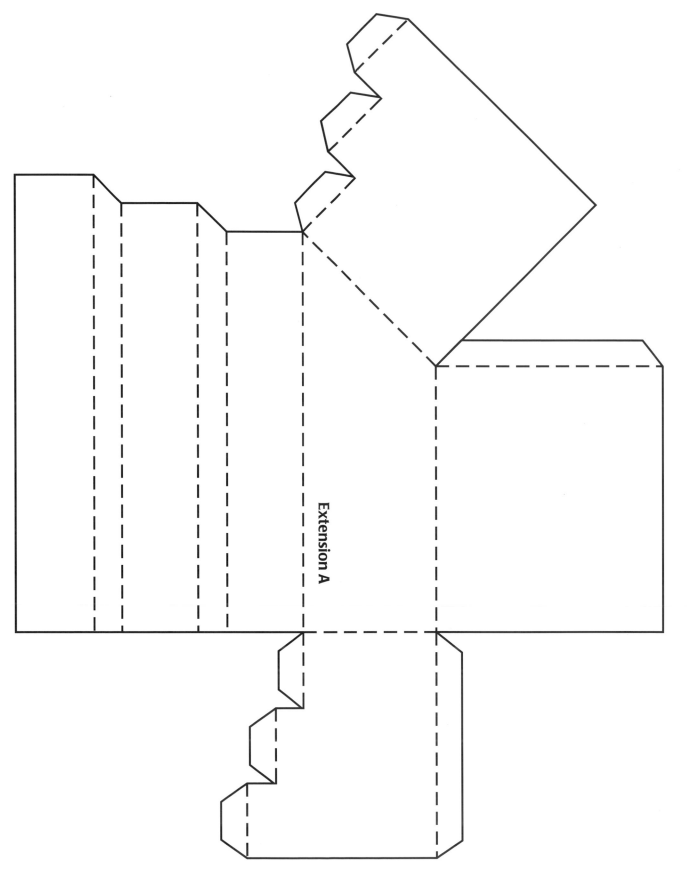

Extension A

© 1998 Encyclopædia Britannica Educational Corporation. This sheet may be reproduced for classroom use.

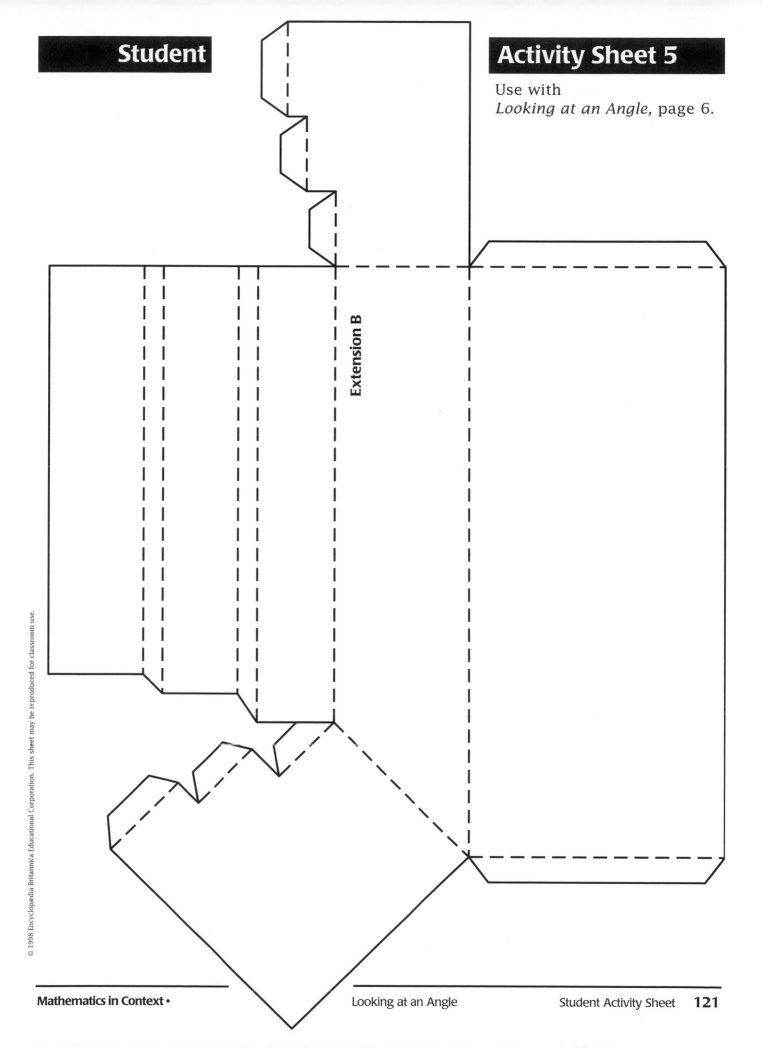

Extension B

© 1998 Encyclopaedia Britannica Educational Corporation. This sheet may be reproduced for classroom use.

16. For each boat shown below, draw a vision line from the captain's position, over the front edge of the boat, to the water. Measure the angle between the vision line and the water. (The captain is located at the star symbol.)

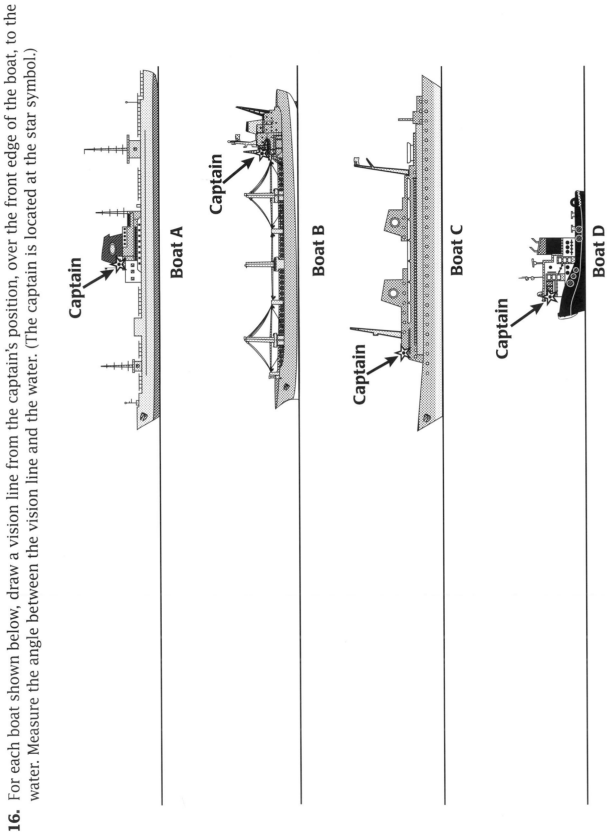

Captain

Boat A

Captain

Boat B

Captain

Boat C

Captain

Boat D

© 1998 Encyclopædia Britannica Educational Corporation. This sheet may be reproduced for classroom use.

© 1998 Encyclopaedia Britannica Educational Corporation. This sheet may be reproduced for classroom use.

Use with *Looking at an Angle,* page 14.

25. a. Draw the captain's vision lines for the side, top, and front views.

  b. In the top view, shade the area of the grid that represents the blind area of the boat.

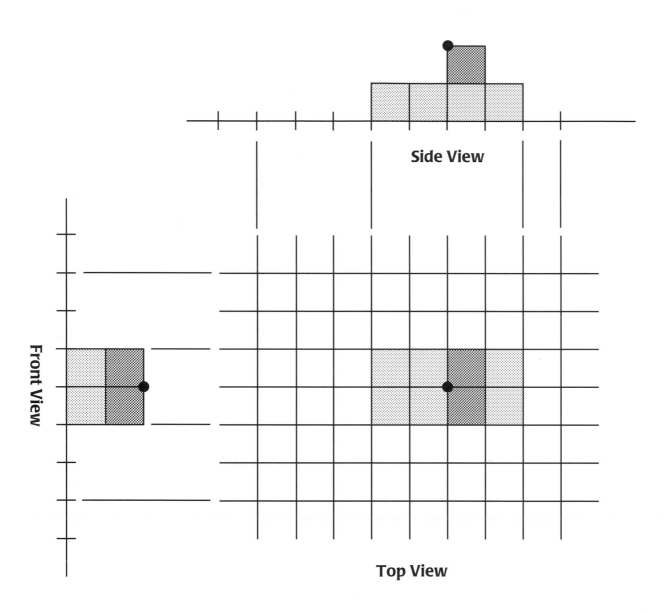

Side View

Front View

Top View

© 1998 Encyclopaedia Britannica Educational Corporation. This sheet may be reproduced for classroom use.

Name_____

26. Draw vision lines and shade the blind area for the view shown below. (One vision line has already been drawn.)

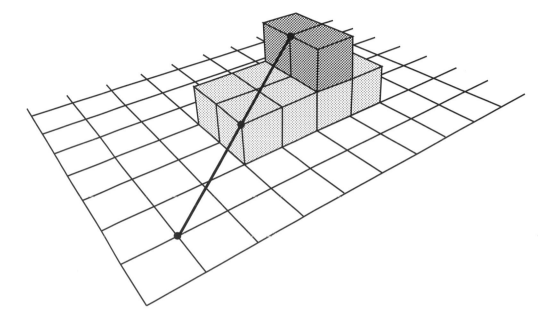

© 1998 Encyclopædia Britannica Educational Corporation. This sheet may be reproduced for classroom use.

6. Draw the shadows of the other tree stumps for each projection.

Nighttime Perspective Projection

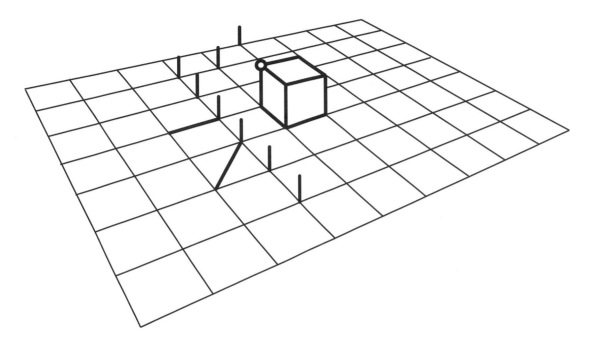

Nighttime Parallel Projection

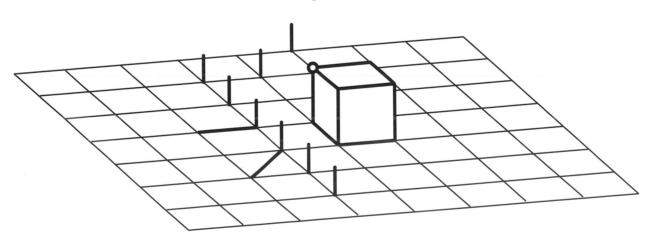

© 1998 Encyclopædia Britannica Educational Corporation. This sheet may be reproduced for classroom use.

Nighttime Top View

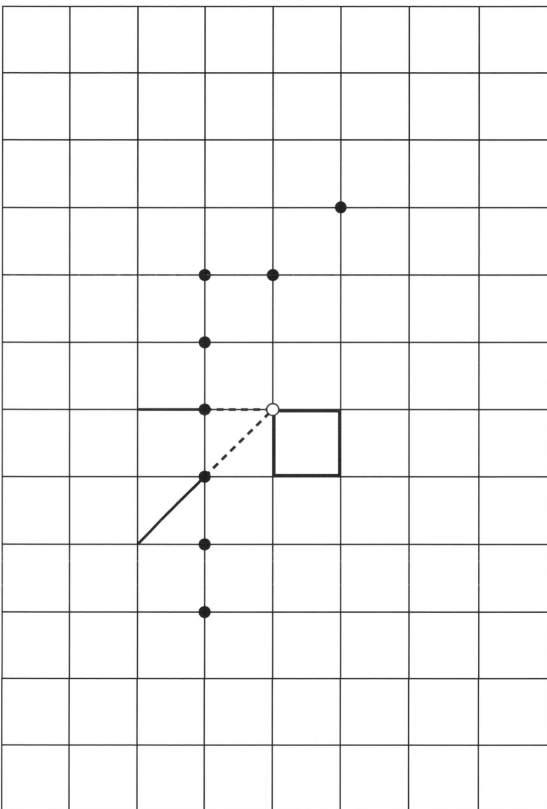

© 1998 Encyclopaedia Britannica Educational Corporation. This sheet may be reproduced for classroom use.

Name _____

Use with *Looking at an Angle,* page 21.

Daytime Perspective Projection

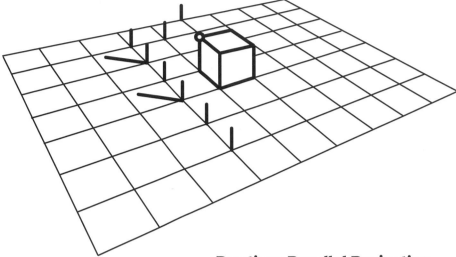

Daytime Parallel Projection

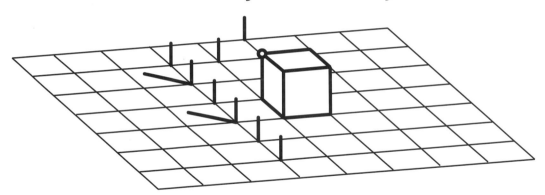

Daytime Top View

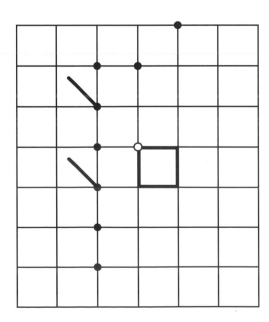

© 1998 Encyclopaedia Britannica Educational Corporation. This sheet may be reproduced for classroom use.

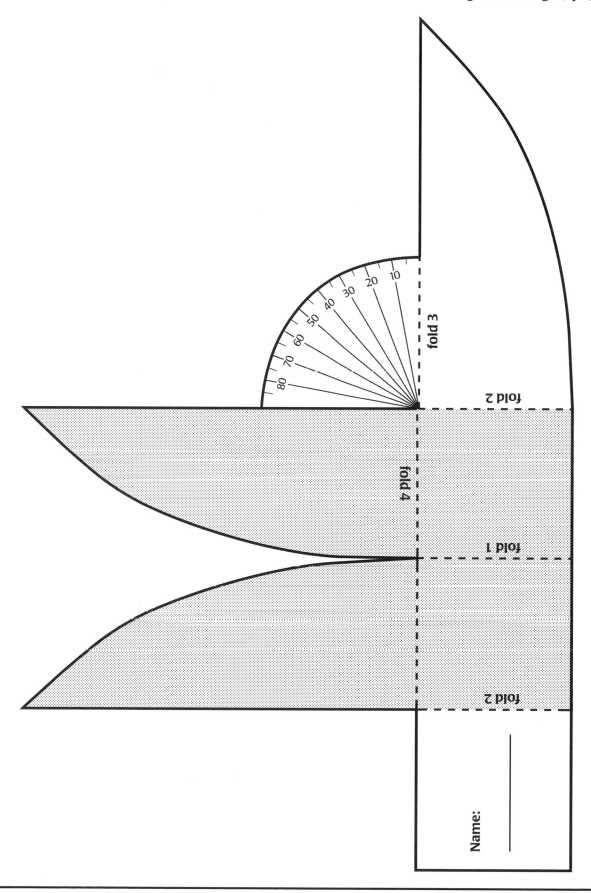

fold 3

fold 2

fold 4

fold 1

fold 2

Name:

© 1998 Encyclopaedia Britannica Educational Corporation. This sheet may be reproduced for classroom use.

Name _____

Use with *Looking at an Angle*, page 25.

17. a. On the following four pictures, draw the shadows that are missing. (*Note:* Picture D needs both shadows drawn in.)

b. Label each picture with an appropriate time of day.

Picture A

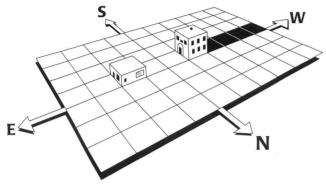

Picture B

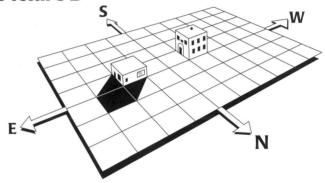

Picture C

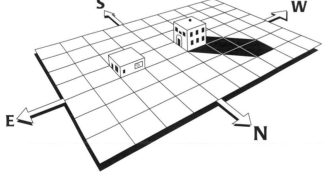

Picture D

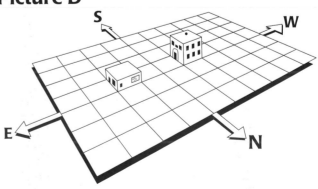

© 1998 Encyclopædia Britannica Educational Corporation. This sheet may be reproduced for classroom use.

23. a. Draw the shadow of the wall.

    b. Does moving the searchlight from one front corner of the fort to the other change the area of the shadow? Explain.

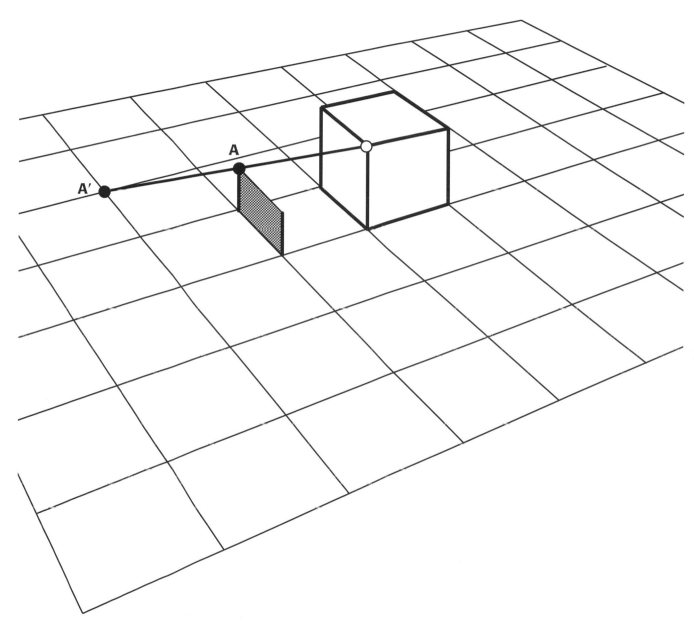

© 1998 Encyclopaedia Britannica Educational Corporation. This sheet may be reproduced for classroom use.

Use with *Looking at an Angle,* page 28.

26. Draw the missing shadows for the following pictures. In top view A, it is nighttime, and the streetlight is on. In top view B, it is daytime, the streetlight is off, and the sun is shining.

Top View A
Shadows from a Streetlight

Top View B
Shadows from the Sun

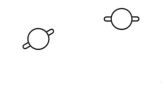

© 1998 Encyclopaedia Britannica Educational Corporation. This sheet may be reproduced for classroom use.

2. On picture A, draw a ray of sunlight that casts a shadow for each of the other 10 rungs.

3. On picture B, draw a ray of sunlight and the corresponding shadow for each of the other 10 rungs.

Picture A

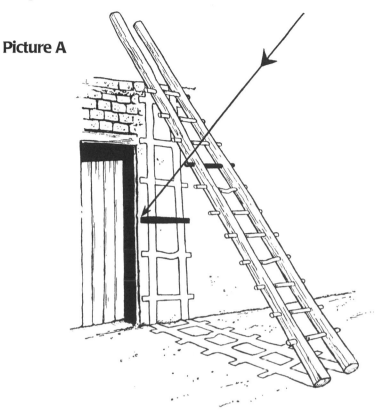

Picture B

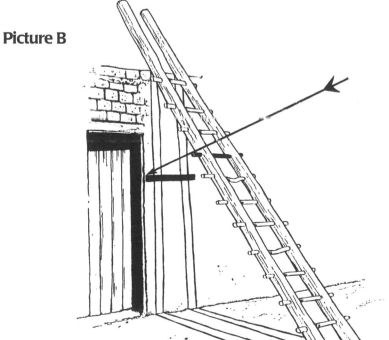

© 1998 Encyclopædia Britannica Educational Corporation. This sheet may be reproduced for classroom use.

Use with *Looking at an Angle,* page 46.

31. A glider with a glide ratio of 1:28 is launched after being pulled by an airplane to 1,200 meters above Lake Havasu City in Arizona. Indicate on the map how far the glider can fly if there is no wind.

Source: © 1997, Encyclopædia Britannica, Inc.

© 1998 Encyclopædia Britannica Educational Corporation. This sheet may be reproduced for classroom use.

Use additional paper as needed.

Two scientists, Jorge and Theresa, are studying in the southwestern part of the United States. They are investigating rock formations, deserts, plants, and animals. They are an adventurous pair—hiking in canyons, climbing rocks, and hang gliding are some of their favorite ways to explore and investigate.

Today they are working in Mammoth Rock Country.

Mammoth Rock Country

Shown below is a map of the area where they are working.

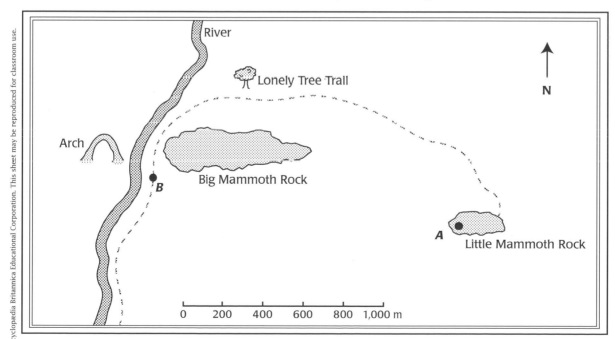

© 1998 Encyclopædia Britannica Educational Corporation. This sheet may be reproduced for classroom use.

Map of Mammoth Rock Country

Use additional paper as needed.

Shown below is a picture that was taken in Mammoth Rock Country.

1. Find a location on the map where the photographer may have been standing while taking this picture. Mark the spot with an "✕" on the map on the previous page.

Jorge and Theresa have just climbed Little Mammoth Rock. Theresa is at point *A* and about to take a hang glider flight. However, she will only start the flight when Jorge has descended the rock and has arrived at point *B*. Jorge is following the trail as he hikes. Theresa watches him from where she is sitting at point *A*. After a while, near Lonely Tree, Jorge disappears behind Big Mammoth Rock. Theresa waits and waits, but Jorge seems to stay out of sight forever. He has been in the "blind area" for half an hour.

2. Outline the blind area behind the rock on the map on the previous page.

3. How long is the part of the trail in that blind area? Make an estimate using the scale line.

Under normal circumstances, a person can walk about 2 or 3 kilometers per hour.

4. Is there a reason for Theresa to be worried? Explain your answer.

© 1998 Encyclopædia Britannica Educational Corporation. This sheet may be reproduced for classroom use.

Name _____ Date _____

Use additional paper as needed.

As Theresa waits for Jorge to reappear, she wonders if he is walking in the shadow of the rock or in the sun. She notices that the shadow of Little Mammoth Rock just reaches the bottom of Big Mammoth Rock. Theresa knows that Little Mammoth Rock is twice the height of Big Mammoth Rock. She draws the shadow on the map. It looks like this:

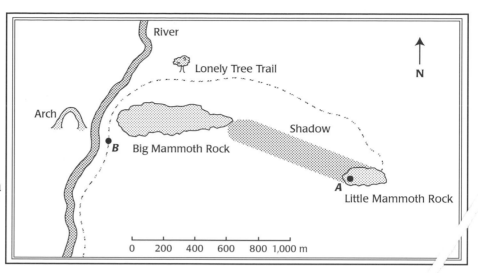

5. Draw the shadow of Big Mammoth Rock in the picture above. Be precise in drawing the shadow's length and direction.

Below you see three top-view drawings of the arch and its shadow at different times on one day.

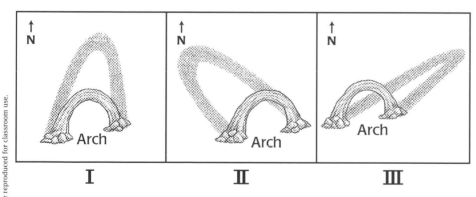

6. In which order were these drawings made?

7. Draw the shadow of the arch at sunset in the picture below. Be precise in drawing the shadow's length and direction.

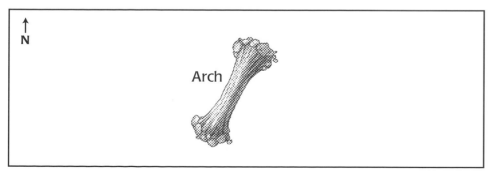

© 1998 Encyclopædia Britannica Educational Corporation. This sheet may be reproduced for classroom use.

Use additional paper as needed.

Below is a sketch of a side view of the two rocks. The line represents the sun's ray. The distance between the two rocks is 910 meters.

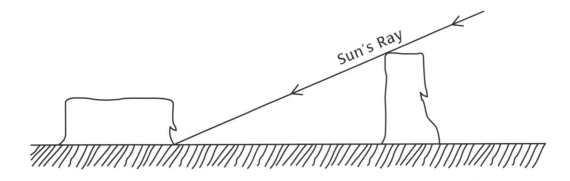

8. Measure the angle between the sun's ray and the horizon with a compass card or protractor.

9. Compute the height of Little Mammoth Rock. Show your work.

Hours later, the rock's shadow is six times longer than the height of the rock.

10. Compute the angle between the sun's ray and the horizon in this situation. Write an explanation of how you found the answer. Show your work.

Finally, Jorge reappears behind Big Mammoth Rock. He uses his walkie-talkie to tell Theresa that an injured hiker has been calling for help. The hiker is on the other side of the river near the arch (see the map). He asks Theresa to fly in her hang glider across the river to the hiker in order to help him. Theresa tells Jorge that she is not sure she can reach that spot. Her hang glider has a glide ratio of 1:6.5.

11. Can Theresa reach the wounded hiker with her hang glider from Little Mammoth Rock? Show the mathematics that supports your answer.

© 1998 Encyclopaedia Britannica Educational Corporation. This sheet may be reproduced for classroom use.

Section A. Now You See It, Now You Don't

1–2.

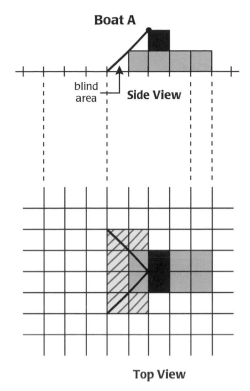

3. The blind area of boat A is 6 square units. The blind area of boat B is 18 square units.

4. The angle between the water and the vision line is 45° for boat A and about 18° for boat B.

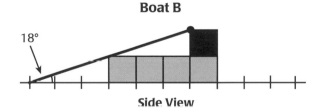

5. The captain of Boat A has the best view. Explanations will vary. Sample explanations:

The captain whose boat has the smallest blind area will have the best view. Boat A's blind area is smaller (6 square units) than boat B's blind area (18 square units). Therefore, the captain of boat A has the best view.

The angle between the water and the vision line for boat A is larger (45°) than the angle for boat B (18°). Boat A has a vision line that is steeper, so its blind area is smaller, and its captain has the better view.

Section B. Shadows and Blind Spots

1. a. The stick's shadow is pointing in the same direction as the pyramid's shadow—northeast.

   b. The sun is shining from the opposite direction—southwest. Shadows produced by the sun point in the direction opposite the position of the sun in the sky.

2. The pyramid is 160 meters tall. Explanations will vary. Sample explanations:

The length of the stick's shadow is 1.5 times the stick's height. Similarly, the length of the pyramid's shadow will be 1.5 times the pyramid's height. Since the pyramid's shadow is 240 meters long, dividing 240 meters by 1.5 gives the height of the pyramid; $240 \div 1.5 = 160$. So, the pyramid is 160 meters high.

There is a fixed ratio between the height of an object and the length of its shadow at a certain moment in time. Using a ratio table shows that the pyramid's shadow is 160 times longer than the stick's shadow. Thus, the height of the pyramid is 160 times that of the stick, or 160 meters.

	Stick	Pyramid
Height of Object (in m)	1	160
Length of Shadow (in m)	1.5	240

× 160

Section C. Shadows and Angles

1. In the sample drawings below, one centimeter represents one meter.

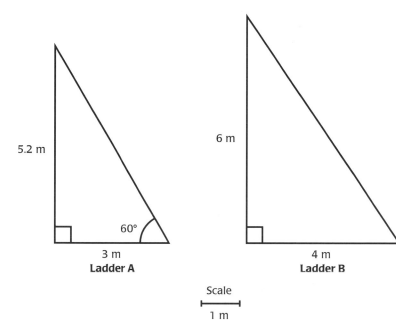

Ladder A (with 5.2 m, 60°, 3 m)

Ladder B (with 6 m, 4 m)

Scale
|—————|
1 m

2. Ladder A: 5.2:3 = 1.7:1

Ladder B: 6:4 = 1.5:1

In order to find the height-to-distance ratio for Ladder A, students need to measure the height of the wall in their scale drawings.

3. about 56°

4. Ladder A is steeper than ladder B. Explanations will vary. Sample explanation:

The steeper ladder will have a larger angle between its foot and the ground. Since the angle between the foot of the ladder and the ground is larger for ladder A (60°) than it is for ladder B (56°), ladder A is steeper than ladder B.

Section D. Glide Angles

1. a. about 3°

   b. about 87°

   (*Note:* Tan A and tan B have opposite *h:d* ratios, $\frac{1}{20}$ and 20. That means that the sum of their angles, 3° + 87°, must be 90°.)

2. The glide ratio for glider II is 1:16. Since the glide ratio of glider I is 1:20, glider I must travel a distance that is 20 times the height of the cliff; 20 × 50 meters = 1,000 meters. The difference between the distance traveled by glider I and that traveled by glider II is 200 meters, so glider II travels 1,000 meters – 200 meters = 800 meters. Therefore, glider II has a glide ratio of 50:800 or 5:80 or 1:16.

3. Side *AB* is about 17.3 centimeters long. The tangent of angle *D* is tan 45° = 1, so the height-to-distance ratio of side *CB* to side *DB* is 1:1. Since side *DB* is 10 centimeters long, that means the length of side *CB* is also 10 centimeters (10:10 is the same ratio as 1:1). Using the table on Student Book page 44, you can find that tan 30° = 0.577, so the height-to-distance ratio of side *CB* to side *AB* can be found by using a calculator and the following ratio table.

	× 1,000	÷ 57.7	
Height of Side *CB* (in cm)	0.577	577	10
Distance of Side *AB* (in cm)	1	1,000	17.3

4. The 30-meter cliff is 1.5 times higher than the 20-meter cliff (20 meters × 1.5 = 30 meters). Because the two gliders have the same glide ratio, the glider launched from the 30-meter cliff will travel a distance that is 1.5 times farther than the distance traveled by the glider launched from the 20-meter cliff.

   To find the location where the gliders land, divide the distance between the two cliffs (100 m) into two parts that are in the ratio 1:1.5. The ratio 1:1.5 is equal to 2:3 = 4:6 = 40:60. Because 40 + 60 = 100, the glider on the left must land 40 meters from the left cliff and the glider on the right must land 60 meters from the right cliff.

5. a. Answers will vary. Sample response:

   A 5% glide ratio means that the hang glider will travel a distance of 20 units when it takes off from a height of 1 unit because 5% = 5/100 = $\frac{1}{20}$, or a height-to-distance ratio of 1:20.

   b. About 3°. The glide angle α for a glider with a glide ratio of 1:20 is tan $\alpha = \frac{1}{20} = 0.05$. Using the table on page 44, students can find that α is equal to about 3°.

Cover

Design by Ralph Paquet/Encyclopædia Britannica Educational Corporation.

Collage by Koorosh Jamalpur/KJ Graphics.

Title Page

Phil Geib/Encyclopædia Britannica Educational Corporation.

Illustrations

8 (top) Phil Geib/Encyclopædia Britannica Educational Corporation; **8 (bottom right)** Paul Tucker/Encyclopædia Britannica Educational Corporation; **10** Phil Geib/Encyclopædia Britannica Educational Corporation; **12** Phil Geib and Paul Tucker/Encyclopædia Britannica Educational Corporation; **16 (top)** Paul Tucker/Encyclopædia Britannica Educational Corporation; **16 (bottom), 20, 26, 28, 30, 32, 34 (bottom)** Phil Geib/Encyclopædia Britannica Educational Corporation; **34 (top)** Paul Tucker/Encyclopædia Britannica Educational Corporation; **36, 40, 42, 50, 52, 54, 58, 60, 62** Phil Geib/Encyclopædia Britannica Educational Corporation; **63** Paul Tucker/ Encyclopædia Britannica Educational Corporation; **66, 68, 70, 76, 80, 84** Phil Geib/Encyclopædia Britannica Educational Corporation; **86** Paul Tucker/ Encyclopædia Britannica Educational Corporation; **88, 90, 92, 94, 96, 98** Phil Geib/Encyclopædia Britannica Educational Corporation; **100** Paul Tucker/Encyclopædia Britannica Educational Corporation; **106, 108, 110, 111** Carrie Schuler/ Navta Associates, Inc.

Photographs

6, 8, 18 © Els Feijs; **22 (top)** © Alastair Black/Tony Stone Images; **22 (bottom)** © Milton & Joan Mann—CAMERAMANN INTERNATIONAL, Ltd.; **24, 34** © Els Feijs; **50** © Jan de Lange; **66** © Els Feijs; **86** © Stock Montage, Inc.

Mathematics in Context is a registered trademark of Encyclopædia Britannica Educational Corporation. Other trademarks are registered trademarks of their respective owners.

CRRC
Math
Second
MIC
Gr 7
v. 8
Teacher
1998

GOSHEN COLLEGE GOOD LIBRARY

3 9310 01013934 1